Radio Occultation Method for Remote Sensing of the Atmosphere and Ionosphere

Edited by
Y.A. Liou

Radio Occultation Method for Remote Sensing of the Atmosphere and Ionosphere

Edited by Y.A. Liou

This is a reprint of the book published in 2010

Published by InTech

Janeza Trdine 9, 51000 Rijeka, Croatia

Technical Editor Goran Bajac

Cover Designer Dino Smrekar

Additional hard copies can be obtained from orders@intechopen.com

Radio Occultation Method for Remote Sensing of the Atmosphere and Ionosphere,
Edited by Y.A. Liou
p. cm.
ISBN 978-953-7619-60-2

For EU product safety concerns:
IN TECH d.o.o., Prolaz Marije Krucifikse Kozulić 3, 51000 Rijeka, Croatia,
info@intechopen.com or visit our website at intechopen.com.

Preface

This book is devoted to presentation of radio occultation (RO) remote sensing as a global method for monitoring of the earth's atmosphere and ionosphere. This technique is based on the following effect: when a spacecrafts radiating radio signals moves into the shadow zone behind the earth and, afterward, appears from this zone, the radio ray produces two cuts of the atmosphere. Atmospheric and ionospheric effects arise in the most cases owing to influence of a zone near the radio ray perigee and cause significant variations of the amplitude, phase, and frequency of the radio waves. These variations enable determination of the altitude profiles of temperature, pressure, refractivity, density, humidity, and turbulence in the atmosphere, distribution of the electron density in the ionosphere, and the wave phenomena at different altitudes with a global coverage.

Aim of this book consists in a systematic description of different approaches, results of investigation, and perspectives of the RO remote sensing as a tool for investigations of the atmosphere and ionosphere. Historical stages of elaboration of RO method, its principle and technical parameters are described in chapter 1. Chapter 2 is devoted to theoretical analysis of effects of radio waves propagation in the communication links satellite-to-satellite. The RO direct problem is stated and analyzed. Variations of the amplitude, phase, and frequency of radio waves relevant to special forms of the altitude profiles of the atmospheric and ionospheric parameters are described. Sensitivity of RO method to variations of the atmospheric temperature, pressure, and electron density in the ionosphere is estimated. Inverse RO problem is discussed and scheme of determination of the altitude profiles of the atmospheric temperature, pressure, refractivity, and electron density in the ionosphere from measurements of the frequency, phase and amplitude is presented. The different radioholographic methods are described in chapter 3: (1) Radioholographic focused synthetic aperture (RHFSA) method; (2) Fourier Integral Operators (FIO) including the Zverev's transform and General Inversion Operator (GIO), (3) Back Propagation (BP) and Canonical Transform (CT) methods; (4) Full Spectrum Inversion (FSI) technique; (5) Spectral Phase Matching Method (SPPM). These methods were elaborated with aim to improve vertical resolution and accuracy in estimation of parameters of the atmosphere and ionosphere and to avoid interfering influence of the multi-path propagation on retrieval of the atmospheric parameters. Also the eikonal acceleration/intensity method is presented and discussed in chapter 3. This technique is useful for identification of layered structures in the atmosphere and ionosphere, evaluation of the intensity of atmospheric and ionospheric irregularities, estimation of the location and parameters of inclined plasma layers in the ionosphere and for excluding of the refractive attenuation from the amplitude data with aim to measure the total atmospheric absorption. Examples of RO signals variations caused by atmospheric influence are adduced in chapter 4, and a step-by-step transfer from RO measurements to determination of the atmospheric parameters is considered. RO measurement errors and inaccuracies of data inversion algorithms influence on the accuracy of retrieved atmospheric parameters. A short description of the basic errors sources is presented in chapter 4. Values of the atmospheric parameters, determined by the RO technique, are compared with the results, obtained by other technical means. RO sounding of the atmosphere allows obtaining information not only about the above mentioned characteristics of the atmosphere, but also about the wave, layered and turbulent structures in the atmosphere, and possibility of their research by the RO method is considered in chapter 4. Influence of the lower ionosphere on the amplitude and phase of RO signal are considered in chapter 5. Physical changes in the near-earth space environment in response to variations in solar radiation, solar plasma ejection, and the electromagnetic status of the interplanetary medium produce disturbances in the ionosphere. The disturbed ionosphere changes the amplitude and phase of RO signal. To the lowest order, changes in the total electron content (TEC) along the signal path contribute to the phase path excess. For an undisturbed ionosphere, where the electron density does not vary significantly over the short- scale lengths, this is the only effect that the ionosphere has on the RO signals. For undisturbed conditions, the tangent points in the ionosphere are absent during motion of the ray perigee in the atmosphere and the ionospheric influence may be described as a slow change (appeared as linear or parabolic trend) in the phase path excess without noticeable variations in the amplitude of RO signal. Analysis of CHAMP data indicates importance of the amplitude variations for classification of the ionospheric influence on RO signals. This classification can be mainly based on the dispersion and on the spectral form of amplitude variations. Strong regular variations in the amplitude of RO signal in the most case are connected with the inclined ionospheric layers. Regular character of the ionospheric disturbances indicates a possibility to obtain additional information about the ionospheric structure from RO measurements. This reveals usefulness of RO method for global investigation of the sporadic E- layers in the lower ionosphere which is difficult to perform by the Earth's based tools.

Two new applications of RO technique are considered in chapter 6: (1) bistatic radio location at small elevation angle and analysis of direct and reflected radio waves propagation effects conducted during MIR/GEO and GPS/MET RO missions at wavelengths 2, 32, 19, and 24 cm; (2) the absorption of centimeter and millimeter radio waves owing to influence of oxygen and water vapor in the troposphere. Experimental observation of propagation effects at low elevation angles has principal importance for fundamental theory of radio waves propagation along the earth's surface. At decimeter wavelength band, the total absorption effect in the trans-atmospheric telecommunication link orbital station MIR – geostationary satellites was measured at frequency 930 MHz. In this experiment, the refractive attenuation has been excluded by use of the phase and Doppler frequency data. Important relationships between the Doppler frequency and the refractive attenuations of the direct and reflected signals are revealed. These connections allow recalculating the Doppler shift to the refractive attenuation and open a possibility to measure the total absorption in the atmosphere by bistatic radar method. GPS/MET and CHAMP (wavelength 19 and 24 cm) RO experiments opened new perspectives for bistatic monitoring of the earth at small elevation angles. The absorption measurements are planning for the future RO missions to determine with high vertical resolution the water vapor abundance at different altitudes in the stratosphere and troposphere. Two directions discussed in chapter 6 broaden the applicable domain of the RO technique.

Y.A. Liou

Radio Occultation Method for Remote Sensing of the Atmosphere and Ionosphere

Y.A. Liou
Center for Space and Remote Sensing Research, National Central University, Chung-Li 320, Taiwan.

A.G. Pavelyev, S.S. Matyugov, O.I. Yakovlev
Institute of Radio Engineering and Electronics of Russian Academy of Sciences (IRE RAS), Fryazino, Vvedenskogo sq. 1, 141190 Moscow region, Russia

J. Wickert
GeoForschungsZentrum Potsdam (GFZ-Potsdam), Telegrafenberg, 14473 Potsdam Germany

The remote sensing satellite radio occultation method elaborated for monitoring of the earth's atmosphere and ionosphere with a global coverage is described. Comparison of theoretical results with experimental observations of radio wave propagation effects in the earth's atmosphere and ionosphere in the communication links satellite-to-satellite is provided. Directions in application of the radio occultation method are discussed: measuring vertical gradients of the refractivity in the atmosphere and electron density in the lower ionosphere, determination of the temperature regime in the stratosphere and troposphere, investigation of the internal wave activity in the atmosphere, and study of the ionospheric disturbances on a global scale. The radio occultation technique may be applied for investigating the relationships between processes in the atmosphere and mesosphere, study of thermal regimes in the intermediate heights of the upper stratosphere-lower mesosphere, and for analysis of influence of space weather phenomena on the lower ionosphere. Radio-holographic methods are considered as a tool for determination of the altitude profiles of temperature, pressure, refractivity, internal wave activity in the atmosphere, and electron density in the ionosphere with usage of the radio links satellite-to-satellite. Results of radio occultation measurements of the atmospheric and ionospheric parameters are described. Comparative analysis of effectiveness of the radio occultation and other remote sensing methods is conducted.

1. Elaboration of Radio Occultation Monitoring of Atmosphere and Ionosphere

1.1 Stages of elaboration of radio occultation method

The RO technique relies on bistatic radio locations when a receiver is located at an extended distance relative to transmitter of radio waves [1]. In distinction with the radio tomography methods (see, for example [2], and references therein), the RO technique may be applied practically simultaneously to investigation of both the atmosphere and ionosphere. The RO technique was initially suggested for remote sensing of planetary atmospheres, ionospheres, and surfaces [1]. During the first space missions to Mars and Venus, a possibility for investigations of their atmospheres and ionospheres by RO technique was used. The RO method is based on the next effect: if a spacecraft immerses into and then egresses from a radioshadow of a planet, a radio ray perigee conducts two «sections» of the planetary atmosphere and ionosphere. According to the atmospheric and ionospheric influence, the regular and irregular variations in the amplitude, phase and frequency of radio waves take place. These variations contain important information about the atmosphere and ionosphere of a planet [1]. The first investigations of the planetary atmosphere by the RO method were conducted during 1965 Mariner-4, 6 and 1969 Mariner-7 Mars flyby's [3,4]. Before interplanetary space flights, Mars investigations were conducted by use of the earth-based spectroscopic observations, which have an inherently large measurement uncertainty in values of the Martian atmospheric pressure and other physical parameters. Information on the Martian ionosphere practically was absent. The RO sounding performed by three Mariner spacecraft has clearly shown that this method makes it possible to determine the pressure and temperature of rarefied atmosphere of Mars and the electron density of Martian ionosphere. In order to employ large informative potential of RO method, artificial satellites of planets have been used. In 1971, massive RO sounding of the rarefied atmosphere and ionosphere of Mars was performed by the first artificial satellites missions to Mars: Mars 2 and Mariner 9 spacecrafts [5, 6].

The first reliable direct measurements of composition, pressure, and temperature in the upper and middle atmosphere of Venus were obtained from USSR entry probe missions. Investigation of Venusian atmosphere via the RO method was started during Mariner 5 and 10 Venus flyby's [7, 8]. Detailed investigations of the atmosphere and ionosphere of Venus started in 1975 with usage of the first Venus artifical satellites Venera 9 and 10. By means of these spacecrafts, the RO experiments at three frequencies were conducted in 50 regions of Venus [9–13]. During these experiments effects of radio waves propagation through the ionosphere and dense Venusian atmosphere were studied. Vertical profiles of temperature $T(h)$ and pressure $P_A(h)$ were obtained independently from measurements of the amplitude and frequency of radio waves. The second series of RO investigations were performed in 1978 by the Pioneer Venus spacecraft [14], and third series of experiments were conducted in 1984 by use of Venera 15 and Venera 16 satellites [15–17]. Investigations of the Venus atmosphere and ionosphere were conducted at the decimeter (λ = 32 cm and 13 cm) and centimeter wavelength bands (λ =8 cm, 5 cm, and 3.6 cm). These multi-frequency measurements allow effective conducting RO investigations of thin atmospheric structures, determining the altitude profiles of temperature, the latitude and longitude distributions of the wind velocities at different altitudes in the atmosphere, detecting the atmospheric turbulence, measuring the altitude profile of sulfuric acid density responsible for the radio waves absorption, and providing detailed study of the ionosphere under different condition of solar illumination. It is important that the RO investigations of the atmosphere and ionosphere were provided in mass scale with global coverage. The first stage of development of the RO method was completed with detailed investigations of the atmospheres and ionospheres of Mars and Venus. A more comprehensive description of this stage is given in [16].

The RO investigations of the earth's atmosphere are possible with usage of two satellites, one of which radiates signals, while the other spacecraft receives them. During motion of the satellites, the radio ray perigee passes through the medium conducting nearly vertical section of the earth's atmosphere at different altitudes. A possibility of RO method application to study the atmosphere and ionosphere of the earth has been considered at the initial stages of investigations. Theoretical estimations of the atmospheric and ionospheric influence on radio waves propagation in the communication link satellite-to-satellite have been provided for revealing a sensitivity of radio waves to features in vertical structures of the atmosphere and ionosphere. Arguments on behalf of RO method in the case of investigation of unknown atmospheres of planets are different from the arguments in the case of investigation of the well-known atmosphere of the earth. In the first case, acquisition of any additional information is justified, while, in the second case, this method should have advantages over the other traditionally ground-based and remote sensing methods for collection of meteorological and ionospheric data. In publications [18–26], problem of the RO remote sensing of the atmosphere and ionosphere of the earth is considered; general relationships for the changes of the frequency, phase, amplitude, bending angle and absorption of radio waves were obtained; estimations of the expected atmospheric and ionospheric effects on radio wave propagation were evaluated for three cases *a*) two satellites are moving at the same orbit supporting nearly the same distance, *b*) geostationary satellite - satellite moving along a low earth orbit (LEO) and *c*) LEO satellite - a satellite of the Global Positioning System (GPS). For these cases, the theoretical dependences of the refractive attenuation, bending angle, variations of the amplitude, frequency and absorption of radio waves were obtained as functions of the altitude of the radio ray perigee. The authors of these publications estimated the necessary accuracies in measurements of the amplitude, frequency, and phase of radio waves with aim to achieve the required precision in determination of the ionospheric and atmospheric parameters including the atmospheric pressure and temperature.

The first RO experiments were made in two satellite-to-satellite links: that of a geostationary satellite and LEO satellite [25] and that of the Apollo-Soyuz Test Project [26]. The RO experiments have shown that the atmosphere and ionosphere change the frequency and amplitude of radio waves in a complex way. Therefore, systematic investigations of the properties of radio wave propagation along the RO satellite-to-satellite paths are required. These investigations were started in Russia in 1990 with the use of the orbital station MIR and two geostationary satellites [27–31]. Radio links of the Ku band (λ = 2 cm) and the UHF radio band (λ = 32 cm) with transmitters of increased power and antennas with high directivity were used. The detailed investigations of the atmospheric and ionospheric influence on the radio waves propagation and estimations of real possibilities of studying the earth's atmosphere and ionosphere by the RO method have been provided by use of these tools in 1990–1998 years. It became evident that the RO system of investigation of atmosphere and ionosphere will be effective when high-stable signals are used. The first studies proposing the usage of highly stable signals of navigational satellites of the GPS and GLONASS systems for sounding the earth's atmosphere and ionosphere appeared in the late 1980-th [23, 24]. A testing RO system was realized in USA in 1995 year with using a LEO satellite Microlab having a receiving device for registration of signals of the navigational satellites GPS, emitting the radio waves in two wavelength bands $\lambda_1 \approx 19$ cm and $\lambda_2 \approx 24$ cm [32–39] . Microlab mission functioned during period 1995 – 1998 years and performed nearly 11 000 measurement sessions. The obtained vertical profiles of the atmospheric temperature and the electron density in the ionosphere were compared with the data of ground-based measurements, and it has been demonstrated that the RO measurements provide a high level of accuracy [32–39].

The second stage of the RO investigations included elaboration of algorithms for the data analysis and practical validation of these algorithms during mission of MIR – geostationary satellites and Microlab – GPS. The second stage was completed with a detailed study of characteristic properties of propagation of decimeter and centimeter radio waves along the satellite-to-satellite paths. As a result of this stage, efficiency of the RO method for exploration of the earth's atmosphere and ionosphere has been demonstrated. It became evident that, in order to provide efficient investigation and monitoring of the atmosphere and ionosphere via the RO method, it is necessary to construct a system that uses several satellite-to-satellite paths simultaneously and to develop new methods for analysis of RO measurements. During the third stage of RO investigation, an international system for global monitoring of the atmosphere and ionosphere was developed (see Table 1.1.1). This system currently included several satellites, which can receive signals from the navigational satellites of GPS system and conduct more than 3000 sessions of RO measurements per day [40–49]. The international RO system uses the satellites – receivers of GPS signals CHAMP (2000), SAC-C (2000), GRACE-A (2002), FORMOSAT-3/COSMIC (2006), METOP (2006), TerraSAR/TanDEM-X (2007), and other, having nearly circular orbits with inclination 75° - 85° at altitudes 500 – 800 km.

Stage	Satellite	Number of satellites	Years of experiments	Country
I	MARS 2	2	1971 – 1972	Russia
	MARINER 9			USA
	VENERA 9 and 10			Russia
	PIONER VENUS	5	1975 – 1984	USA
	VENERA 15 and 16			Russia
II	MIR– GEOSTATIONAR	2	1990 – 1998	Russia
	MICROLAB 1	1	1995 – 1998	USA
III	CHAMP	1	2001	Germany
	GRACE	2	2002	Germany – USA
	FORMOSAT-3/COSMIC	6	2006	Taiwan –USA
	Metop-A	1	2006	ESA
	TerraSAR-X	1	2007	Germany

Table 1.1.1. Stages of elaborating of RO method for remote sensing of the atmosphere and ionosphere

1.2. RO system for monitoring the atmosphere

To obtain information about the atmosphere and ionosphere for meteorology, climatology, and geophysics, it is required (1) a global coverage of the earth's surface by the RO measurements; (2) high accuracy of measurements and usage of radio signals in different frequency bands. Global sounding may be fulfilled only by use of many satellites, transmitting radio waves, and satellites - receivers of signals. The time period required for sounding of the atmosphere in a given region should be essentially shorter than the time scale corresponding to the changes in the atmospheric state, and the frequency of measurements in any region should correspond to the frequency of observations usual for standard meteorological practice, i.e. one time per six hours. A system consisting of high orbital satellites with long orbital period and satellites installed in low orbits satisfies these requirements because difference in orbital periods the low orbital satellites will periodically immerse into or egress from the earth's limb relative to the high orbital satellites, providing RO sounding of the atmosphere above different regions. The scheme of RO sounding of the atmosphere is shown in Fig. 1.2.1. In Fig. 1.2.1 the satellites, transmitting the radio waves, are located in points G_j, j=1,...n+2, and the satellite-receiver of signals is disposed at point L, point T corresponds to the radio ray perigee and is disposed at the minimal altitude above the earth surface. For supporting this system of remote sounding of the atmosphere, the navigational satellites are used as emitters of radio waves. This solves the problem of global coverage of the earth and assures high accuracy in measurements of atmospheric parameters owing to high stability of signals, emitted by navigational satellites.

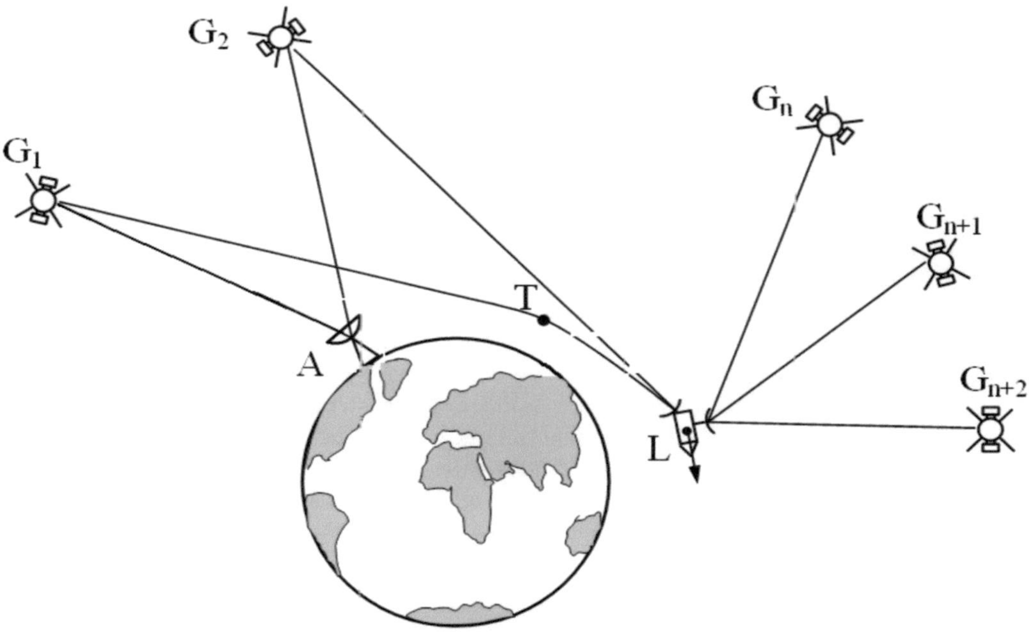

Fig. 1.2.1. Scheme of RO remote sensing. G_1 is the occulted navigational satellite, G_2 is the reference satellite, L is the low orbital satellite-receiver, G_n... G_{n+2} are satellites for measuring the orbital parameters of the low orbital satellites, and A is the ground-based station for receiving RO information and data analysis.

A detailed description of the stages of elaboration and designing of navigational systems, basic principles, and structure of emitted signals and usage for RO investigations are published in [50–52]. Navigational satellite systems are destined for solution of navigational problems, i.e. for determination of the coordinates and velocities of different objects on the surface of land and sea, in the atmosphere and in the earth's environmental space. Coordinates and velocity of any object may be determined from the results of measurements of its distance from any three navigational satellites, and velocity – from the changes of these distances, i.e. from the radial velocity. In the radio-technical systems, the distances, as a rule, are determined from the signal delay, and the radial velocity – from the Doppler shift of its frequency. To increase the accuracy of measurements of the signal delay, it is necessary to broaden its spectrum. To increase the accuracy of measurements of the Doppler shift, it is necessary, vice versa, to increase the signal duration. This contradiction may be avoided under a condition of joint estimation of the delay and Doppler shift in the case of application of signals with large base. The signal base is equal to the product of its duration and effective spectral width. Application of the noise-like signals with large base is necessary for functioning of the navigational system. In the navigational satellite systems GLONASS and GPS for achieving the high resolution and stability relative to the noise and interferences, the noise-like signals with phase-manipulation are applied. These signals consist of the impulse sequences with initial phases having discrete values 0 and π. The initial impulse signal with duration τ is divided to N elements. Each of these elements has duration $\tau_N = \tau/N$. In this case, the equivalent spectral width of the noise-like signal is by a factor $B = \tau / \tau_N$ greater, than the same one of the initial signal.

Navigational system GPS consists of ~29 satellites (canonically 24 plus a few spares) that are distributed in six circular orbital planes, having inclination 55° to the equatorial plane. The angle between the orbital planes is equal to 60°. The altitude of satellites is equal to 20,180 km, orbital period is about 11 h 58 m. Distribution of the satellite on the orbits assures observing five or more satellites above any region of the earth's surface. Each GPS satellite continuously broadcasts signals in two frequency bands L1 and L2. All GPS satellites transmit signals at the same carrier frequencies, which are formed from the frequency f_0=10.23 MHz. The carrier frequency f_1 (band L1) is equal to 154 f_0, and the carrier frequency f_2 (band L2) – 120 f_0, i.e. f_1=1575.42 MHz, and f_2=1227.6 MHz. The ratio of the carrier frequencies is equal to f_2/f_1=60/77. Signals in bands L1 and L2 are coherent and modulated by the two pseudo-random codes: the basic P-code with the speed of transmission 10.23 Mb/s and open C/A code with the speed of transmission 1.023 Mb/s. Diagrams of transmitting antenna illuminate practically uniformly the earth's hemisphere, as seen from the satellite. The power of the GPS signal at the output of a linearly polarized ground based antenna having the gain coefficient +3 dB, is greater than –163 dBW for channel L1 when using P-code, or –160 dBW for C/A code, and –166 dBW for channel L2. It is planning from 2014 year, that system GPS will have satellites of new generation with increased values of the signal power.

For supporting the global navigational radio field the navigational system GLONASS will have 24 satellites, orbiting around the Earth in three planes. The orbits of the GLONASS satellites are near circular with the altitude about 19,100 km, orbital period 11 h 15 m 44 s and inclination 64.8°. Orbital planes are displaced by 120° on longitude of the ascending node. In each orbital plane, eight satellites are disposed with 45° latitude shifting, the satellites in the neighboring orbital planes are displaced by 15°. This structure of the

GLONASS constellation assures observation in any region of the earth. Four or more GLONASS satellites are continuously transmitting the coherent signals in the two bands L1 and L2. The carrier frequencies in the bands L1 and L2 are formed coherently from the reference frequency 5 MHz. The ratio of carrier frequencies, emitted by a separate satellite GLONASS in the bands L2 and L1, is equal to f_2/f_1=7/9. The GLONASS satellites are transmitting the navigational signals of the standard and heighten accuracy. The signals of the standard accuracy are formed by modulation of the carriers f_1 and f_2 with the frequency 0.511 MHz, the heighten accuracy signals are modulated by a special code with a chip rate 5.11 MHz. The power of the GLONASS signals at the output of a linearly polarized ground based receiving antenna having the gain coefficient +3 dB, is greater than –161 dBW for the frequency band L1, and –167 dBW for the frequency band L2

The spaceborne and ground-based segments constitute the system of RO monitoring of the atmosphere. The spaceborne segment includes the navigational satellites (point G in Fig. 1.2.1) and several satellites - receivers in the low orbits (point L), having qual-frequency receiver and antenna for the navigational and RO measurements. A key element of the satellite L is a measuring receiver, conducting registration of the amplitude, phase path excess of radio waves, and coordinates for navigation. Navigational measurements are conducted with the sampling frequency 0.1 Hz by use of antenna with zenith orientation. One or two directional antennae are installed for the RO measurements on the satellite L. If one directional antenna is installed on a satellite L, the axis of its diagram is located in the orbital plane of the satellite and oriented to the earth's limb in the direction, opposite to vector of orbital velocity. This antenna assures sounding the atmosphere during setting of the receiving satellite behind the earth's atmosphere relative to a navigational satellite. Antenna, oriented in direction of the orbital motion of the receiving satellite, is destined for observation of the rising navigational satellites. Installation of two antennae increases by about two times a number of regions, sounded at one orbital turn of a satellite.

For supporting necessary altitude resolution in determination of the atmospheric and ionospheric characteristics measurements of the signal parameters should be conducted with a high sampling frequency. this requires a special onboard memory device for storage of the results of measurements before their transmission to an earth-based receiver station. To diminish the required volume of memory, the sounding of the upper ionosphere is provided with the small sampling frequency (10 Hz), and when the minimal altitude of radio ray G_1L is achieved 130 km measurements with the large sampling frequency (50 or 100 Hz) are provided. The results of measurements, concentrated in the onboard memory device, are periodically transmitted to an earth-based receiving stations, and then to the Center of Guidence and Data Analysis. The structure of the ground based part of the system contains a net of the stations for receiving of the satellite information, the measuring centers that control the orbits and the time onboard the navigational satellites, and the Center for Guidance and Data Analysis. The scheme shown in Fig. 1.2.1 demonstrates the different variants of RO sounding: «autonomous», when measurements are conducted only in one communication link G_1L, «differential» for measurements with usage of two communication links G_1L and G_2L, and the technique double differencing measurements, that used an ground-based center (point A).

The Center for Guidance and Data Analysis provides analysis of all information for supporting high accuracy in the retrieved parameters of a sounded medium and forms the data bank having several levels and comparison of the onboard clocks of all navigational (G) and measuring satellites– receivers (L) and their correction to the precise atomic clock. Also determination of the coordinates and vectors of the satellites velocities, estimations of the quality of the primary data, the data analysis with usage of different methods for obtaining the altitude profiles of the parameters of the sounded medium are conducted in the Center for Guidance and Data Analysis. The results of measurements and data analysis are retained, as a rule, in three-level format. The first level contains detailed information describing measurement conditions. The second level, accessible for customers, contains the date and time of beginning of measurement session, session number, number of a navigational satellite, taking part in RO sounding, number of channel of measuring receiver, the time interval between the sequential samples, value of signal-to-noise ratio in the band L1 and L2, coordinates and component of velocities of the navigational and LEO satellites, and values of the phase path excess of the signals L1 and L2, caused by influence of a medium. Analysis of these data allows one to determine the parameters of a medium in different region of the earth. Values of the physical parameters of a medium, determined from the results of analysis of the RO signals, are listed in the data of the third level. These data contains date, time and number of measurement session, altitude in the atmosphere, latitude and longitude of the investigated region, refractivity, air density, pressure and temperature in the atmosphere, the bending angle, and other parameters.

In the present time, the satellites GRACE-A, GRACE-B, METOP-A, FORMOSAT-3/COSMIC, and TerraSAR - X are used for the RO sounding of the atmosphere and ionosphere. These satellites provide more than 3000 measurement sessions per day. Experimental information received from these satellites is analyzed in real time in the centers of data analysis in USA, Germany, and Taiwan. This system provides the global control of the current state of the atmosphere and ionosphere and allows one to solve the next problems:

- monitoring the altitude distribution of temperature, density, and pressure with high accuracy and high vertical resolution for improvement of weather prediction and studying the climate changes;
- providing control of the geopotential altitude;
- monitoring of the distribution of water vapor for better understanding the role of the global water vapor circulation in the meteorological and climatological problems;
- control of the turbulence and internal atmospheric waves distribution;
- monitoring the ionosphere and revealing connection of the ionosphere and upper atmosphere with the solar activity and antropogenic influence.

2. Direct and Inverse Problems of Radio Occultation Remote Sensing

2.1. Refractivity, rays, and bending angle

A direct problems of RO investigation of the atmosphere or ionosphere is resolved to determine the changes of the amplitude, phase or frequency of radio waves in the communication link satellite-to-satellite, if vertical profile of the refractivity $N(h)$ is known. This problem was investigated in detail in the publications [16, 21, 22, 28, 29, 37, 48, 52-59]. We will follow these publications during analysis of the RO direct problem. Geometry of the RO direct problem is shown in Fig. 2.1.1. The satellites are disposed in the points L and G at the altitudes H_l and H_g, the earth's center is located at point O, the radio ray LTG has in the point T a minimal altitude H above the earth surface. The radio ray in the free space, in the segments LL_1 and GG_1, is a straight line, in the segment L_1G_1 the ray is curved because of the medium influence. Change in the ray direction is described by the bending angle ξ. Let us assume that the atmosphere or ionosphere is a locally spherical symmetric medium. Hence, on the ray segment L_1G_1 near the point T, one may neglect influence of the horizontal gradients of medium and the refractivity index $n(r)$ depends only on the distance $OC = r = a + h$. Let us introduce the altitude h of the arbitrary point C on the ray trajectory and designate a is the earth's radius, θ is the central angle LOG, $r_g = a + H_g$, $r_l = a + H_l$, and $r_t = a + H$ are, correspondingly, the distances OG, OL, and OT. The decimeter and centimeter radio waves are used for the RO sounding so that the medium parameters insignificantly change at a distance equal to the wavelength. Therefore, one may apply the geometric optics for the analysis of the direct problem. For a spherically symmetric medium, the following relationships are valid

$$n(r)\, r\, \sin\gamma = \text{const}\,, \qquad (2.1.1)$$

$$P\, \Delta S = \text{const}\,, \qquad (2.1.2)$$

where γ is the angle between the radius–vector $\mathbf{r}$ and the unit vector of a radio ray $\boldsymbol{I_0}$.

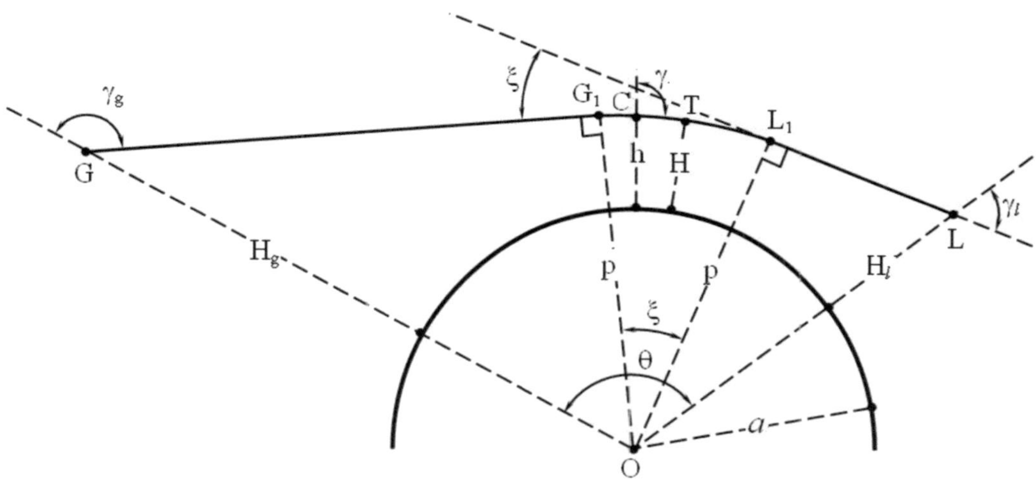

Fig. 2.1.1. Geometry of the RO direct problem.

Equation (2.1.2) connects the density of the power flow P and the cross section of a ray tube ΔS. Eq. (2.1.2) allows determining the changes in value P, caused by refraction of radio waves. Derivation of the relationships (2.1.1) and (2.1.2) is described in many monographs on radio waves propagation (see for example, [59]). It follows from (2.1.1) that the altitude dependence of the refractive index $n(r)$ determines the main features of radio waves propagation.

Let us consider the features of the altitude profile $n(h)$. The distinction of the refractive index $n(h)$ from unity is very small. Therefore, it is convenient to introduce the refractivity N equal to $N = n - 1$. The refractivity N depends on the pressure P_a, temperature T_a, and humidity w_a

$$N = \frac{77.6}{T_a}\left(P_a + \frac{4810\, w_a}{T_a}\right)\cdot 10^{-6}, \tag{2.1.3}$$

where the pressure and humidity are expressed in millibars, and the temperature is measured in Kelvin. The hydrometeors in the troposphere (rain, snow, mist, ... etc) introduce a small addition ΔN, determined by an approximate relationship $\Delta N \approx 1.4\, w_w$ where w_w is the water content expressed in g/m^{-3}. It is important that N in the atmosphere does not depend on the frequency. In the troposphere, the pressure and humidity are diminishing when the altitude h is increasing according to an exponential dependence, and the temperature has a nearly linear change as function of height. Hence, the altitude profile of the refractivity may be approximated by an exponential law

$$N \approx N_0 \exp\left(-b_1\, h\right), \tag{2.1.4}$$

where N_0 is the refractivity near the earth surface. Values N_0 may be determined by use of Eq. (2.1.3) from measurements of P_a, T_a, and w_a. In the moderate latitudes, N_0 on average is equal to $3.06 \cdot 10^{-4}$ in winter, in summer this value is near to $3.3 \cdot 10^{-4}$; the parameter b_1 is equal to 0.13 km^{-1}, it changes in 0.12 – 0.14 km^{-1} interval. Value b_1 may be determined from the magnitude of N_0 if one accounts for insignificant changes of N at the altitude 10 km, where N is equal to $9.2 \cdot 10^{-5}$. Therefore, with accounting for expression (2.1.4), one may obtain

$$b_1 = -\frac{1}{10} \ln\left(\frac{9.2 \cdot 10^{-5}}{N_0}\right). \tag{2.1.5}$$

It follows from the relationships (2.1.3), (2.1.4), and (2.1.5), that the approximated dependence $N(h)$ may be found from the near surface values of pressure, temperature, and humidity. The real profiles $N(h)$ may differ from the exponential dependence. The more accurate form of vertical profile $N(h)$ may be described by the relationship

$$N(h) = N_0 \exp\left(-a_1\, h^2 - b_1\, h\right). \tag{2.1.6}$$

Approximation (2.1.6) corresponds in average good to the actual dependence $N(h)$, however it does not account for the features of $N(h)$ at the tropopause and in the troposphere, where temperature inversions and clouds permanently exist . More detailed information on the altitude distribution of N in the troposphere is given in [60–62].

Let us consider dependence of the refractive index on the frequency and altitude in the ionosphere. It is known that the plasma's refractivity is determined at high frequencies by a simple relationship

$$N = -\chi N_e f^{-2} , \tag{2.1.7}$$

where $\chi = 40.3$, the electron density N_e is expressed in m^{-3}, and f if the frequency in Hz. Derivation of this formula is given, for example, in [59]. It follows from (2.1.7) that N is negative and dependence $N(h)$ repeats the altitude profile of the electron density of ionosphere $N_e(h)$. The refractivity diminishes as f^{-2} when the frequency f increases. Vertical profile $N_e(h)$ may be described by different ways in the area, located above the main ionospheric maximum, when $h>h_m$, and in the lower part of the ionosphere when $h<h_m$. In the upper part of the ionosphere $N_e(h)$ may be satisfactorily described by an exponential dependence

$$N_e = N_m \exp\left[-b_2\left(h-h_m\right)\right], \tag{2.1.8}$$

where N_m is the electron density in the main ionospheric maximum, h_m is the altitude of the main maximum of the electron density, b_2 is the parameter, characterizing the speed of diminishing of the electron density when the altitude increases. For the part of the ionosphere, located below the main ionospheric maximum, it is difficult to find the appropriate approximation, describing dependence $N_e(h)$. As a rough approximation in this region, one may use a formula

$$N_e \approx N_m \left[1-\left(\frac{h_m-h}{d_2}\right)^2\right]. \tag{2.1.9}$$

This approximation corresponds to abatement of the electron density up to zero at the altitude $h = h_m - d_2$ where d_2 is the approximated width of the lower part of the ionosphere. Vertical profiles $N_e(h)$ depends on the daytime, season, latitude, and solar activity. If $h<h_m$, dependence $N_e(h)$ has a complex form: in this area the regular ionospheric layers F_1 and E are located, irregularly additional sporadic plasma E_s layers appear in this area. The rough approximation (2.1.9) does not account for these features. There are numerious publications with detailed description of the distribution of electron density. The International Reference Model of the Ionosphere (IRI) [63, 64] has been designed for presentation of the standard altitude dependences $N_e(h)$ in the Internet.

Let us analyze the refraction of radio waves in situation shown in Fig. 2.1.1. It follows from formula (2.1.1)

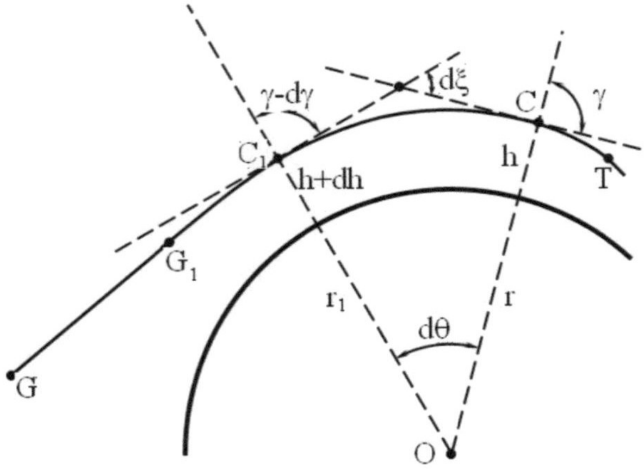

Fig. 2.1.2. Geometrical scheme for estimation of the refractive attenuation.

$$(a+H)n(H)=(a+H_g)n(H_g)\sin\gamma_g=(a+H_l)n(H_l)\sin\gamma_l= \\ =(a+h)n(h)\sin\gamma=p. \tag{2.1.10}$$

We assume that $n(H_g)=n(H_l)=1$, therefore

$$(a+H)n(H)=(a+H_g)\sin\gamma_g=(a+H_l)\sin\gamma_l=(a+h)n(h)\sin\gamma=p\,.$$

In these relationships $p=L_1O=G_1O$ is the impact distance (or the impact parameter beeing constant on the radio ray), *H* is the minimal altitude of radio ray in point T, *n*(*H*) is the refractive index in this point. The radius-vector **r** and radio ray in the points G, L, and C constitute the angles γ_g, γ_l and γ. The angles LL_1O and GG_1O are equal to 90° (Fig. 2.1.1). From Eq. (2.1.10) one has a relationship for the ray in a spherically symmetric medium

$$\text{tg}\,\gamma=\frac{p}{\left(r^2n^2(r)-p^2\right)^{1/2}}\,. \tag{2.1.11}$$

It follows from Eqs. (2.1.10) and (2.1.11), that a radio ray passing through the given points L and G is determined by the altitude profiles of the refractive index $n(h)$ and parameter *p*. In the case of multi-path propagation through point L and G, several ray lines with different values of *p* and *H* may pass.

Below we will find a relationship for the curvature radius of a radio ray in a spherically symmetric medium $R_0=dl/d\xi$. We consider two points C and C_1 on a ray. The corresponding change of the bending angle is $d\xi$ and the length element is $dl = CC_1$. Then, according to the geometrical scheme shown in Fig. 2.1.2, one can obtain the next relationships

$$d\xi = d\gamma + d\theta \ , \tag{2.1.12}$$

$$dh = dl\cos\gamma \ , \tag{2.1.13}$$

$$d\theta = (a+h)^{-1}\,\mathrm{tg}\gamma dh \ , \tag{2.1.14}$$

where $d\theta$ is the angle between the vectors $\mathbf{r}_1$ and $\mathbf{r}$, $d\gamma = \gamma - \gamma_1$, $dh = h_1 - h$. Accounting for Eqs. (2.1.12), (2.1.13), and (2.1.14), one may obtain the following relationship for radius of the ray curvature

$$R_0 = \frac{a+h}{\sin\gamma + (a+h)\dfrac{d\gamma}{dh}\cos\gamma} . \tag{2.1.15}$$

By usage of Eq. (2.1.10), one may obtain

$$R_0 = \frac{-n}{\dfrac{dn}{dh}\sin\gamma} , \tag{2.1.16}$$

where n, γ and R_o depend on the altitude h. It follows from (2.1.16), that in the point T

$$R_0 = -n\left(\frac{dn}{dh}\right)^{-1} .$$

Below we will find the bending angle altitude dependence. According to (2.1.12), the bending angle is equal to

$$\xi = 2\int\limits_H^\infty \left(\frac{d\gamma}{dh} + \frac{d\theta}{dh}\right) dh , \tag{2.1.17}$$

where the factor 2 accounts for the refraction along the lines LT and GT. By use of Eqs. (2.1.14) and (2.1.10), one may find

$$\xi = -2\int\limits_H^\infty \frac{1}{n}\frac{dn}{dh}\,\mathrm{tg}\gamma dh . \tag{2.1.18}$$

By use of Eqs. (2.1.10) and (2.1.18), one may obtain a formula for the bending angle in a spherical symmetric medium

$$\xi(p) = -\,2p \int\limits_{\mathrm{p}}^{\infty} \frac{1}{n}\,\frac{dn}{dh}\left(r^2n^2 - p^2\right)^{-1/2} \mathrm{dr} , \tag{2.1.19}$$

where, according to (2.1.10), $p = (a+H)n(\mathrm{H})$. Below we assume that, according to (2.1.4) and (2.1.8), $N(h)$ depends on the altitude as an exponential function. Then one may obtain a simple approximation for the altitude dependence $\xi(H)$. After introducting a new variable $x = bh$ and assuming that the next inequality is fulfilled $ba\,N(H) << 1$, one may obtain from (2.1.19)

$$\xi = 2b\,(a+H)\,N_0 \int_{bH}^{\infty} \frac{\exp(-x)\,dx}{\left[x\,\left(x+2b\,(a+H)\right)\right]^{1/2}}, \qquad (2.1.20)$$

therefore

$$\xi = 2b\,\left(a+H\right)\,N_0\,\exp\left(-ba\right)\,J_0\left(b\,\left(a+H\right)\right), \qquad (2.1.21)$$

where J_0 is the Bessel function of an imaginary argument, $N(H)$ is the refractivity at the altitude H. Assuming that $b(a+H) >> 1$ and by use of the asymptotic presentation of J_0, one has

$$\xi \approx N_0\left(2\pi\,ba\right)^{1/2}\exp\left(-bH\right). \qquad (2.1.22)$$

It follows from Eq. (2.1.22) that, if the refractivity $N(h)$ depends on the altitude according to an exponential law, then the bending angle will depend on the minimal altitude of radio ray H according to the same law. By use of the approximated dependence $N(h)$ (2.1.6), one may determine from (2.1.19) the bending angle for different atmospheric conditions and for different regions. Typical values of the bending angle ξ are in the 70-84, 380-390, 1410-1500, and 3400-3900 intervals of angular seconds for the altitudes H equal to 29, 20, 10, and 2 km, respectively. In the troposphere (H= 1-7 km), the bending angle may have strong variations. The atmospheric refraction does not depend on the wavelength, and in the ionosphere the bending angle ξ is proportional to the square of the wavelength. Refraction in the ionosphere depends on vertical gradient of the electron density according to (2.1.7):

$$dn/dh = -\chi\,f^{-2}\,\frac{dN_e}{dh}. \qquad (2.1.23)$$

In the upper ionosphere under conditions $h > h_m$, and $h < h_m$, when the approximation (2.1.8) is valid, the radio ray is deflected correspondingly from or to the earth surface.

2.2. Refractive attenuation, frequency changes and phase of radio waves

Refractive effect leads to deformation of rays structure. Let us consider a ray tube, having at point G in the plane of Fig. 2.2.1 the angular size $d\gamma_g$, and in the perpendicular plane the dihedral angle $d\chi$, and then find the size of the ray tube in point L. This tube is bounded in the plane of Fig. 2.2.1 by dotted ray lines GL and GL_2. A circle having the radius r_l and center in point O intersects with the dotted ray lines in points L, L_2, therefore $LL_2 = rd\theta$. The linear size of the ray tube in point L is equal to

$$LL_3 = LL_2\cos\gamma_l = r_l\cos\gamma_l d\theta .$$

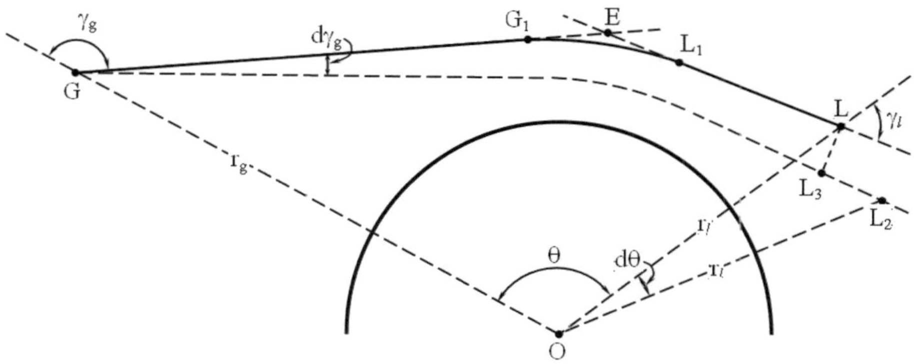

Fig. 2.2.1. Evolution of the ray tube owing to refraction effect in a spherical symmetric medium.

The size of the ray tube in the plane perpendicular to the plane of Fig. 2.2.1 is equal to $r_l \sin\theta \, d\chi$. The cross section of the ray tube S_1 in point L is equal to

$$S_1 = r_l^2 \sin\theta \, \cos\gamma_l \, d\theta \, d\chi \, . \tag{2.2.1}$$

In the free space when the refraction effect is absent the ray tube having in the point G the angular sizes $d\gamma_g$ and $d\chi$, will have in point L the cross section

$$S_0 = L^2 \sin\gamma_g d\gamma_g d\chi \, ,$$

where $L^2 = r_l^2 + r_g^{\;2} - 2\, r_l \, r_g \cos\theta$ is the distance between the point L and G. The refractive attenuation of radio waves X is equal to a ratio of the power flow in the case when the refraction is present P_1 to the power flow in the case of radio waves propagation in free space P_0

$$X = \frac{P_1}{P_0} = \frac{S_0}{S_1} = \frac{L^2 \sin\gamma_g d\gamma_g}{r_l^2 \sin\theta \cos\gamma_l d\theta} \, . \tag{2.2.2}$$

To exclude from transformation (2.2.2) the angles γ_g and γ_l, one may introduce the impact distance p and refraction indexes n_g and n_l relevant to the points G and L. To achieve this, one may use a connection (2.1.10)

$$d\gamma_g = \left(n_g^2 r_g^2 - p^2\right)^{-1/2} dp \tag{2.2.3}$$

$$\sin\gamma_g = \frac{p}{n_g r_g} \, , \qquad \cos\gamma_l = \left(1 - \frac{p^2}{n_l^2 r_l^2}\right)^{1/2}$$

Let us introduce the derivative $d\theta / dp$ in the relationship (2.2.2) instead $d\theta / d\gamma_g$. To achieve this aim consider a quadrangle LOGE , where the point E in Fig. 2.2.1 corresponds to intersection of the tangents to the ray lines LL_1 and GG_1 . From this quadrangle, it follows

$$\theta = \gamma_l - \gamma_g + \xi . \tag{2.2.4}$$

After accounting for (2.1.10), one may obtain

$$\theta = \sin^{-1}\left(\frac{p}{n_g\, r_g}\right) - \sin^{-1}\left(\frac{p}{n_l\, r_l}\right) + \xi \tag{2.2.5}$$

and after differentiation of equation (2.2.5) with respect to p , it follows

$$\frac{d\theta}{dp} = \frac{d\xi}{dp} - \left(n_g^2\, r_g^2 - p^2\right)^{-1/2} - \left(n_l^2\, r_l^2 - p^2\right)^{-1/2} . \tag{2.2.6}$$

After substitution of (2.2.3) and (2.2.6) in (2.2.2), one may find a relationship for the refractive attenuation of radio waves

$$X = \frac{p\left(r_l^2 + r_g^2 - 2\, r_l\, r_g \cos\theta\right)}{r_l\, r_g \sin\theta\left[\left(r_l^2 - p^2\right)^{1/2} + \left(r_g^2 - p^2\right)^{1/2} - \frac{d\xi}{dp}\left(r_l^2 - p^2\right)^{1/2}\left(r_g^2 - p^2\right)^{1/2}\right]} . \tag{2.2.7}$$

It is assumed during elaboration of Eq. (2.2.7) that $n_g = n_l = 1$. The final relationship (2.2.7) allows one to analyze the refractive attenuation of radio waves for a general case of arbitrary dependence of the refractivity on the altitude. If the bending angle is small (this case may correspond to the upper part of the atmosphere and to high frequencies in the ionosphere), then the formula (2.2.7) may be simplified by use of the approximate relationships

$$\begin{gathered} r_l^2 + r_g^2 - 2\, r_l r_g \cos\theta = L^2 \approx \left(L_g + L_l\right)^2 , \\ r_l^2 - p^2 = L_l^2 , \quad r_g^2 - p^2 = L_g^2 , \end{gathered} \tag{2.2.8}$$

where $L_g \approx \mathrm{GE}$, $L_l \approx \mathrm{LE}$, and the distance between the satellites $\mathrm{GL} = L$. It follows from Eq. (2.2.7), with accounting for (2.2.8)

$$X = \frac{p\left(L_g + L_l\right)^2}{r_g\, r_l \sin\theta\left(L_g + L_l - L_g L_l \frac{d\xi}{dp}\right)} . \tag{2.2.9}$$

According to (2.2.9), the refractive attenuation is determined by distances L_g, L_l and derivative $d\xi / dp$. Eq. (2.2.9) allows finding dependence of the refractive attenuation on the altitude H for arbitrary vertical profile $n(h)$. In the case, when $L_g >> L_l$ one may obtain from Eq. (2.2.9)

$$X \approx \frac{p}{r_l \sin\theta \left(1 + L_l \dfrac{d\xi}{dp}\right)}, \tag{2.2.10}$$

When the bending angle is small, $p \approx r_l \sin\theta$, it follows from (2.2.10)

$$X \approx \left(1 - L_l \frac{d\xi}{dp}\right)^{-1}. \tag{2.2.11}$$

For the exponential dependence of the refractivity $N(h)$ on the altitude (2.1.22), one may obtain from (2.2.11) a simplified relationship

$$X \approx \left[1 + \mathrm{b}L_l \left(2\pi\ \mathrm{b}\ a\right)^{1/2} N_0 \exp\ \left(-\ \mathrm{b}H\right)\right]^{-1}. \tag{2.2.12}$$

The relationship (2.2.12) gives an estimation of X for $H > 8$ km in the case when the relationship (2.1.4) is a good approximation of the real dependence $n(h)$. In the interval $H \le 8$ km, where the layered structures exist, it is necessary to use Eq. (2.2.9), determining the derivative $d\xi/dp$ by use of more realistic dependence $n(h)$. In publications [28, 29], dependence $X(H)$ for different profiles $N(h)$ was analyzed. This analysis demonstrated, that the refractive attenuation X is notable at the altitude $H \approx 25$ km, and at the height $H = 20$ km $X = 2$ dB. At the altitudes $H = 18$-25 km, dependence $X(H)$ has a good correspondence with the approximate formula (2.2.12), because at these altitudes, where the atmosphere is nearly isotermic, the formula (2.1.4) is valid. At the tropopause ($H = 9$-17 km), there are significant changes of $X(H)$ with sharp maximum and minimum. The amplitude of these variations depends on the specifical features of the temperature altitude profile $T(h)$. In the troposphere, owing to influence of layered structures, the altitude dependence $X(H)$ has strong variations, in average at the heights $H = 8$ and 2 km, the refractive attenuation is equal to 5.5 and 8 dB, respectively. The ionospheric changes of the amplitude of decimeter waves may be significant in the altitude interval $H \approx 100$–150 km because influence of a large vertical gradient of the electron density. The ionospheric refractive attenuation may be strong in the meter wavelength band so that the case of critical refraction can be observed.

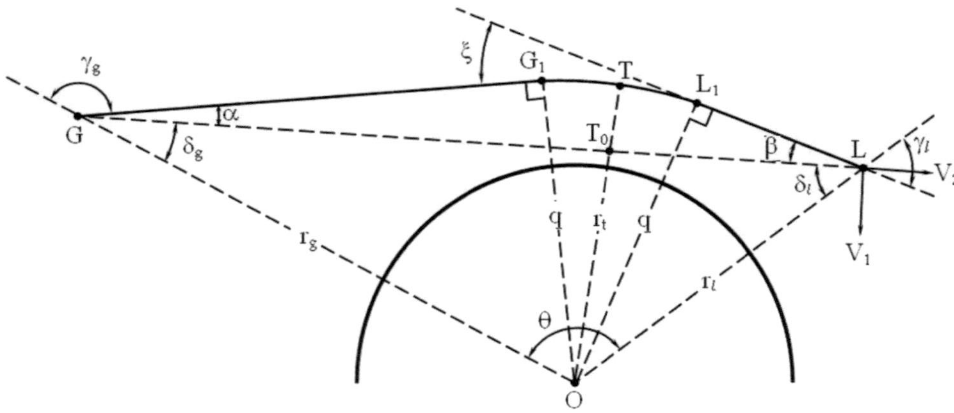

Fig. 2.2.2. Geometric parameters used for estimation of the Doppler frequency.

To find the changes of frequency, caused by an influence of the atmosphere or ionosphere, consider the scheme in Fig. 2.2.2. The coordinate system is connected with location of the GPS satellite. In this system, the satellite G is moveless, the satellite L is moving with the projection v_1 of the velocity vector in the plane of Fig. 2.2.2 on the perpendicular to the dotted straight line GL, and v_2 is projection of the satellite L velocity on the straight line GL. Let us introduce the angles α and β between the tangents to the ray lines in the point G and L and the straight line GL. The Doppler shift due to the atmospheric or ionospheric influence Δf_s is determined by projection v_1 and v_2 on the ray line at point L

$$\Delta f_s = \lambda^{-1}\left(V_2 \cos\beta + V_1 \sin\beta\right). \tag{2.2.13}$$

When the atmosphere is absent, the frequency change due to the satellite motion is equal to

$$\Delta f_0 = \lambda^{-1}\, V_2. \tag{2.2.14}$$

The frequency change Δf due to only the atmospheric or ionospheric influence is determined by the difference

$$\Delta f = \Delta f_s - \Delta f_0 = \lambda^{-1}\left[V_2\left(\cos\,\beta - 1\right) + V_1 \sin\beta\right]. \tag{2.2.15}$$

Therefore, the frequency change due to atmospheric or ionospheric influence is determined by the angle β and components v_1, v_2 of the satellite velocity. From (2.2.15), one has

$$\beta = \sin^{-1}\left[\frac{V_1\left(\lambda\Delta f + V_2\right) - V_2\left(V_1^2 - \lambda^2\Delta f^2 - 2\lambda V_2 \Delta f\right)^{1/2}}{V_1^2 + V_2^2}\right]. \tag{2.2.16}$$

The angle β depends on the bending angle ξ since from the geometrical scheme shown in Fig. 2.2.2 it follows

$$\begin{aligned}
L^2 &= r_g^2 + r_l^2 - 2\, r_g r_l \cos\theta, \\
\delta_g &= \sin^{-1}\left(r_g L^{-1} \sin\theta\right), \\
\delta_l &= \sin^{-1}\left(r_l\, L^{-1} \sin\theta\right), \\
p &= r_l \sin^{-1}\left(\beta + \delta_l\right), \\
\alpha &= \sin^{-1}\left(p\, r_g^{-1}\right) - \delta_g, \\
\xi &= \alpha + \beta.
\end{aligned} \tag{2.2.17}$$

Formula (2.2.15) or (2.2.16) and relationships (2.2.17) give a connection of the Doppler shift Δf and bending angle ξ for a general case, when r_l and r_g are arbitrary. If the satellites have the same altitudes and $r_l = r_g$, then $\xi = 2\beta$; if $r_g >> r_l$, then $\xi \approx \beta$. If the bending angle is small and $r_g >> r_l$, then a simple approximation follows from Eq. (2.2.15)

$$\Delta f \approx \lambda^{-1}\, V_1\, \xi\,. \tag{2.2.18}$$

According to Eq. (2.2.15) or (2.2.18), the frequency change Δf owing to the atmospheric or ionospheric influence is inversely proportional to the wave length and is depending mainly on the bending angle ξ and the satellite velocity component V_1. If V_1 = 1.5 km·s^{-1} and λ = 30 cm, the typical atmospheric frequency change is about 3.6 Hz at the 20 km altitude. When the altitude H diminishes, the frequency Δf increases according to a nearly exponential law and achieves 80-110 Hz at the altitude $H = 4$ km. According to changes of the meteorological conditions and vertical profiles $N(h)$ variations in the altitude dependence of the bending angle ξ and frequency shift $\Delta f(H)$ take place. In the ionosphere, according to (2.1.7) and (2.1.19), $\xi \sim f^{-2}$. Therefore, the ionospheric changes in the frequency shift Δf are raising when the wave length increases. In the lower ionosphere, Δf is relatively small: for V_1 = 1.5 km·s^{-1} and λ = 30 cm Δf changes in the - 0.5…+ 1.5 Hz interval in the $H \approx 80\ldots120$ km altitudes interval.

Let us analyze the atmospheric changes of the phase $\Delta\varphi = \varphi - \varphi_0$, where φ is the signal phase relevant to the curved ray GTL, and φ_0 is the signal phase, corresponding to the dotted straight line GL in the case when the atmosphere is absent (Fig. 2.2.2). The signal phase φ is determined by an integral

$$\varphi = k \int_G^L n\, dl = k \int_G^L \frac{n\, dr}{\cos\psi} = k \int_{r_t}^{r_g} \frac{n^2 r\, dr}{\left(n^2 r^2 - p^2\right)^{1/2}} + k \int_{r_t}^{r_l} \frac{n^2 r\, dr}{\left(n^2 r^2 - p^2\right)^{1/2}}\,. \tag{2.2.19}$$

This relationship follows from the geometrical scheme Fig. 2.2.2 and Eqs. (2.1.10), because of connections $dr = \cos\psi \; dl$, $rn \sin\gamma = p$, $k = 2\pi \; \lambda^{-1}$. If the atmosphere is absent, then

$$\varphi_0 = k \left[\left(r_g^{\,2} - p_0^{\,2}\right)^{1/2} + \left(r_l^2 - p_0^{\,2}\right)^{1/2} \right], \tag{2.2.20}$$

where $p_0 = \mathrm{T_0O}$ is the minimal distance from the straight line GL to the point O. After integration on part in Eq. (2.2.19), one can find

$$\varphi = k \left[\left(n_g^{\,2} r_g^{\,2} - p^2\right)^{1/2} + \left(n_l^{\,2} r_l^2 - p^2\right)^{1/2} - \int_{r_t}^{r_g} \frac{nr^2 \frac{dn}{dr} dr}{\left(n^2 r^2 - p^2\right)^{1/2}} - \int_{r_t}^{r_l} \frac{nr^2 \frac{dn}{dr} dr}{\left(n^2 r^2 - p^2\right)^{1/2}} \right]. \tag{2.2.21}$$

Eq. (2.2.21) may be transformed to the relationship

$$\varphi = k \left[\left(n_g^{\,2} r_g^{\,2} - p^2\right)^{1/2} + \left(n_l^{\,2} r_l^2 - p^2\right)^{1/2} - \right. \\ \left. - \int_{r_t}^{r_g} \frac{1}{n} \frac{dn}{dr} \left(n^2 r^2 - p^2\right)^{1/2} dr - \int_{r_t}^{r_l} \frac{1}{n} \frac{dn}{dr} \left(n^2 r^2 - p^2\right)^{1/2} dr + p\xi \right]. \tag{2.2.22}$$

Assume that in the points G and L the refractivity indexes are equal to 1, $n_g = n_l = 1$. Then, one can find the phase difference from Eqs. (2.2.22) and (2.2.21)

$$\Delta\varphi = k\left(L_1^* + L_2^*\right) = k\left[\left(r_g^2 - p^2\right)^{1/2} + \left(r_l^2 - p^2\right)^{1/2} - \left(r_g^2 - p_0^2\right)^{1/2} - \left(r_l^2 - p_0^2\right)^{1/2} + p\xi\right] + \\ + k\left[2\int_{r_t}^{\infty} \frac{1}{n}\frac{dn}{dr}\left(n^2 r^2 - p^2\right)^{1/2} dr\right]. \tag{2.2.23}$$

where $p = L_1O = G_1O$, and $p\xi = \mathrm{L_1G_1}$. The first term in the square brackets (2.2.23) corresponds to the difference between the distance along the curved ray $\mathrm{GG_1L_1L}$ and the distance along the straight line GL. The second term (2.2.23) is relevant to the changes in the phase path because of vertical gradients of the refractivity. The phase difference in (2.2.23) is expressed by use of eikonal $\Delta\;\varphi = 2\pi\lambda^{-1}(\mathrm{L}^*_1 + \mathrm{L}^*_2)$. The atmospheric change $\Delta\varphi$ is inversely proportional to the wave length, the ionospheric part of $\Delta\varphi$ is proportional to the wave length. Eikonal $L^*_1 + L^*_2$ has approximately values 1.9 m, 15 m, 97 m and 464 m for the altitude H equal to 29, 20, 10 and 2 km, respectively.

The refractive attenuation X, Doppler shift Δf and phase difference $\Delta\varphi$ may be expressed as functions of the bending angle ξ. Therefore dependences $X(H)$, $\Delta f(H)$ and $\Delta\varphi(\mathrm{H})$

are connected. When finding approximate connections we assumed for simplicity that the bending angle is small, the distance OG is essentially greater than the distance OL, i.e. $L_g >> L_l$, and accounted for a relationship

$$\Delta f = \frac{1}{2\pi}\frac{d\ (\Delta\varphi)}{dt} \tag{2.2.24}$$

by usage of Eqs. (2.2.11), (2.2.18) and (2.2.24), one can obtain the next relationships

$$X \approx \left[1-\left(\frac{L_l c}{f\ V_1^2}\right)\cdot\frac{d(\Delta f)}{dt}\right]^{-1}, \tag{2.2.25}$$

$$X \approx \left[1-\frac{L_l c}{2\pi f\ V_1^2}\cdot\frac{d^2(\Delta\varphi)}{dt^2}\right]^{-1}. \tag{2.2.26}$$

These approximate Eqs. are obtained under assumption that v_1 does not depend on time. It follows from Eqs. (2.2.25) and (2.2.26), that variations of the refractive attenuation are proportional to changes of derivative of the frequency shift Δf with respect to time or to the acceleration of the phase difference $\Delta\varphi$. These approximate relationships may be useful for identification of contributions from layered structures or for observation of wave phenomena in the atmosphere.

2.3. Inverse problem of RO remote sensing

Solution of the inverse RO problem consists in finding the altitude profiles $N(h)$ and $N_e(h)$ or other parameters of medium from experimental dependences $\Delta\varphi(t)$, $\Delta f(t)$, and $X(t)$. For solution of inverse problem the satellite coordinates and their velocities will be considered as known functions of time. It follows from Eqs. (2.2.7) or (2.2.9) that the refractive attenuation X is determined by the derivative $d\xi / dp$. The frequency shift Δf, according to Eq. (2.2.15) or (2.2.18), depends on ξ. It is essential, that $d\xi / dp$ may be determined from the amplitude, and ξ – from the independent frequency or phase experimental data. To solve inverse problem it is necessary to find the altitude profile $N(h)$ from dependences $d\xi / dp$ or ξ on time by use of Eq. (2.1.19). Let us introduce the new variables

$$y = r^2 n^2 - R^2\ , \qquad z = p^2 - R^2\ , \tag{2.3.1}$$

where R is the radius, corresponding to the upper boundary of the atmosphere or ionosphere. Then from Eqs. (2.1.19) and (2.3.1) one may obtain

$$\xi = -2p \int_z^0 (y-z)^{-1/2} \frac{d[\ln n(r)]}{dy}\, dy\ . \tag{2.3.2}$$

The Abel transformation should be applied to solve Eq. (2.3.2)

$$F_1(z) = -\int_0^z (y-z)^{-1/2} \frac{d F_2(y)}{dy} dy ,$$
$$F_2(y) = \frac{1}{\pi} \int_0^y (z-y)^{-1/2} F_1(z)\, dz , \qquad (2.3.3)$$

where $F_{1,2}$ are the arbitrary functions. It follows from Eqs. (2.3.3) that solution of Eq. (2.3.2) has a form

$$\ln[n(r)] = \frac{1}{\pi} \int_{rn}^{\infty} \xi(p)\left(p^2 - r^2 n^2\right)^{-1/2} dp . \qquad (2.3.4)$$

Eq. (2.3.4) gives inverse transformation of the bending angle ξ to the altitude profile of the refractive index of radio waves $n(r)$. Eq. (2.3.4) may be simplified accounting for the next approximation $\ln[n(r)] \approx N(r)$, $rn \approx r$:

$$N(r) = \frac{1}{\pi} \int_{r}^{\infty} \xi(p)\left(p^2 - r^2\right)^{-1/2} dp . \qquad (2.3.5)$$

This relationship determines the altitude profile of the refractivity $N(h)$, if the bending angle ξ is found from the frequency shift Δf or the phase difference $\Delta\varphi$. Depending on the construction the low orbital satellite receiver can conduct the frequency measurements Δf (frequency meter) or the phase difference $\Delta\varphi$ (phase meter). If the phase difference $\Delta\varphi\,(t)$ is measured then for usage of Eq. (2.2.24) one should differentiate $\Delta\varphi\,(t)$ with aim to find Δf . The refractive attenuation of radio waves may be applied for solution of the RO inverse problem. The following connection between p , $X(t)$ and the impact parameter of the straight line GL p_s has been found in [22]

$$pdp / dt = X(t)\, p_s dp_s / dt \quad . \qquad (2.3.6)$$

Solution of Eq. (2.3.6) can be given in an integral form [22]

$$p - p_0 = \int_{p_{s0}}^{p_s} X[t(p_s)] dp_s \quad . \qquad (2.3.7)$$

Eq. (2.3.7) gives a temporal dependence of the impact parameter $p(t)$ if the initial values $p_0 = p_0(t_0)$ and $p_{s0} = p_{s0}(t_0)$ at the time instant t_0 are known. Assuming $|p-p_s| << p_s$, one obtains from (2.2.11)

$$\frac{d\xi(p)}{dp}=\left[1-\frac{1}{X(t)}\right]\frac{L}{L_g L_l}. \quad (2.3.8)$$

Equation (2.3.8) gives temporal dependence of $d\xi(p)/dp$. Thus the amplitude data may be used for restoring the impact parameter $p(t)$, and the derivative $d\xi(p)/dp$. These temporal dependencies may be used to find vertical distribution of vertical gradient of the refractivity $dN(h)/dh$ [65, 66]

$$dN(h)/dh=-n^2(h)J(p)/\{p[1+J(p)]\},$$
$$J(p)=\frac{1}{\pi}\int_p^{\infty}\frac{d\xi}{dx}\ln\left(\frac{x}{p}+\sqrt{\frac{x^2}{p^2}-1}\right)dx. \quad (2.3.9)$$

Eqs. (2.3.8), (2.3.9) can be applied for solution of the RO direct and inverse problems: finding the refractive attenuation variations for given refractivity profile (direct problem) and retrieving the refractivity distribution from given refractive attenuation variations.
Vertical profiles of the electron density of the ionosphere $N_e(h)$ may be found also from formula (2.1.7).
Accounting for additional relationships the atmospheric temperature $T_a(h)$ and pressure $P_a(h)$ can be found. To determine the atmospheric temperature T_a it is necessary to use the ideal gas state equation

$$P_a = K_b\, n_a\, T_a\,, \quad (2.3.10)$$

where K_b =1.381 10^{-23} Дж/K is the Boltzmann constant, equation of hydrostatic equilibrium

$$dP_a=-m_a g\, n_a dh\,, \quad (2.3.11)$$

where m_a is the molecular mass, g is the gravitational acceleration, n_a is the gas density, and connection between the refractivity and temperature, pressure, and gas density

$$N=\mu_1 P_a T_a^{-1}=\mu_1 K_b n_a\,. \quad (2.3.12)$$

Factor μ_1 depends on the gas composition. For the dry air, when influence of humidity may be neglected, according to Eq. (2.1.3), $\mu_1=77.6$. By usage of the system of Eqs. (2.3.9)–(2.3.12), one may obtain

$$T_a(h)=\frac{T_1 N_1}{N(h)}-\frac{m_a}{K_b N(h)}\int_{h_1}^{h} g(h)N(h)dh\,, \quad (2.3.13)$$

where T_1 and N_1 are the temperature and the refractivity at the altitude h_1. To find T_a(h) from (2.3.13) it is necessary to introduce the initial values of parameters T_1 and N_1 at the boundary of the upper part of the atmosphere, for example at the height h_1=50 km. The inaccuracy in the initial values of T_1 and N_1 practically does not influence on vertical profile of temperature below h<35 km.

As shown in [66] for the case of a wet atmosphere a connection exists between vertical gradients of the temperature and refractivity

$$\frac{dT_w(h)}{dh}[T_w(h)]^{-1} = -[N(h)]^{-1}\frac{dN(h)}{dh} - \frac{T_x}{T_a(h)}, \tag{2.3.14}$$

$$T_w(h) = T(h)/\{1+4810e(h)/[P(h)T(h)]\}, \quad T_x \approx 34.16 \text{ K/km}, \tag{2.3.15}$$

$$T_a(h) = T(h)[1+0.378e(h)/P(h)], \tag{2.3.16}$$

where $T(h)$ is the temperature of the atmosphere in K, $T_w(h)$ is the "wet" temperature of the atmosphere depending on the water vapor pressure $e(h)$ and atmospheric pressure $P(h)$ in hPa, respectively. Eqs. (2.3.14)–(2.3.16) connect vertical gradient of the logarithm of the refractivity with Vertical gradient of logarithm of the "wet" temperature $T_w(h)$. At the height above 10 km Eqs. (2.3.14)–(2.3.16) may be used to find vertical gradient of the temperature profile if the refractivity gradient is known. Integration of Eq. (2.3.14) may give vertical profile $T_w(h)$ if an initial condition is known at some height h_∞

$$T_w(h) = T_w(h_\infty)N(h_\infty)N^{-1}(h) + \\ +T_x N^{-1}(h)\int_h^{h_\infty} N(h)/\{(1+0.378e/p)[1+4810e/(pT)]\}dh. \tag{2.3.17}$$

Eq. (2.3.17) gives vertical temperature profile $T_w(h)$ for the general case of the wet atmosphere. For the case of dry atmosphere $e(h)$=0 and more simple equation may be obtained from Eq. (2.3.17) [67]

$$T(h) = T_x N^{-1}(h)\int_h^\infty N(x)dx. \tag{2.3.18}$$

Practically solution of the RO inverse problem uses multi-stage algorithm of the experimental data analysis. The scheme of monitoring of the atmosphere and ionosphere by RO method is given in the Table 2.3.1.

Block 1 includes input experimental data containing changes of the amplitude, phase or frequency of radio waves. Dependence of the impact distance on time $p(t)$ is obtained from the satellite trajectory data (block 2) and relationships (2.1.10). Then dependence ξ or $d\xi/dp$ on the impact distance p may be found by use of expressions (2.2.15) and (2.2.9) (block 3). Block 4 corresponds to the Abel transformation in form (2.3.5) or (2.3.6). The Abel transformation allows finding the altitude profile of the refractivity from dependence of the bending angle on the impact distance. Electron density of the ionosphere is found in block 10 according to Eq. (2.1.7) from data of block 4, and then from formula (2.3.12) the gas density is determined (block 6). Altitude dependences of temperature and pressure are found from Eq. (2.3.13) (block 9) and Eq. (2.3.9) (block 8), respectively. As follows from this table the practical solution of the inverse RO problem is very complex. The first stages including accurate registration of variations of the amplitude, phase or frequency of radio waves as functions of time and their connection with the trajectory data are especially difficult. The described approach does not account for the influence of humidity this is valid for altitude $h > 8\ km$. For accurate temperature determination in the troposphere it is necessary according to (2.1.3) account for the contribution of humidity w_a in the refractivity N. The altitude distribution of humidity $w_a(h)$ should be found from additional data, for example, from absorption of radio waves (block 13). Detailed determination of dependence $w_a(h)$ from RO data is given in section 4. Variations of the amplitude and phase of radio waves give additional information about atmospheric waves and small-scale irregularities in the atmosphere and ionosphere (blocks 12 and 11) . Method, corresponding to blocks 12 and 11, will be described in chapter 4.

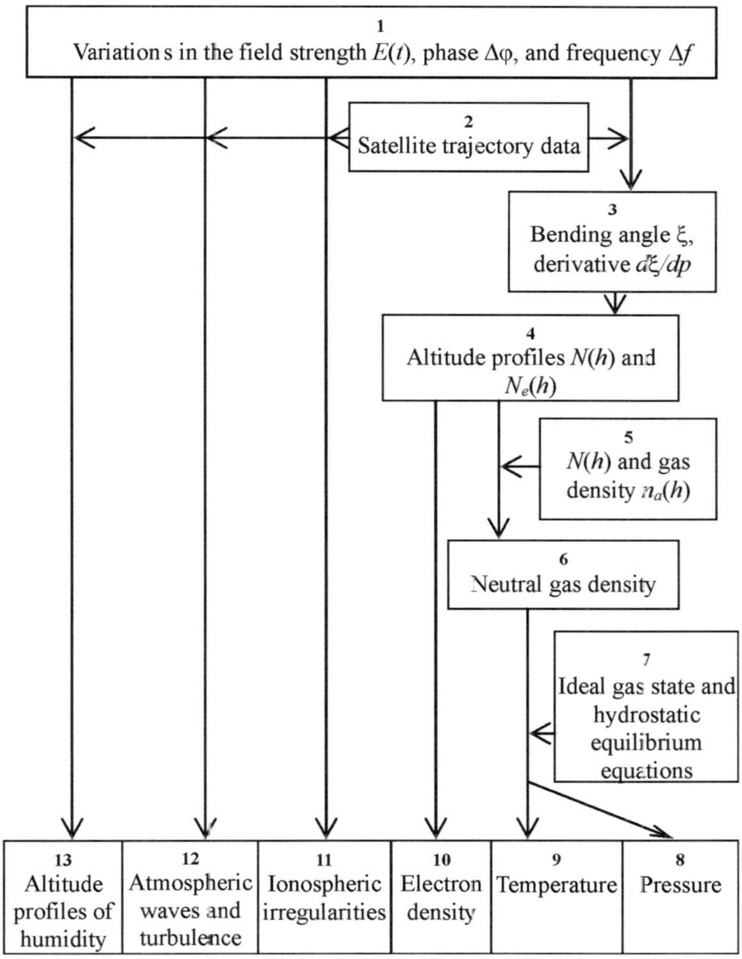

Table 2.3.1. Scheme of determination of the atmospheric and ionospheric parameters by radio occultation method

Two essential assumptions have been made in this section: (1) medium is spherically symmetric and (2) multi-path propagation is absent. Also reflections of radio waves by the earth surface should be accounted for because the satellite antenna diagrams are broad and reflected radio waves are registered in the receiver. However by appropriate filtration of the received RO signal the influence of the reflected wave may be excluded.

3. Radio-Holographic Methods For Solution Of The RO Inverse Problem

3.1. Radio waves, rays and back-propagation. Problem statement

In the case of one ray propagation, the orbital trajectory is intersected by only one ray that has the bending angle ξ and the Doppler frequency Δf. The bending angle ξ can be found if the ray direction is known. However, it is difficult to measure the ray direction near the caustics and multi-path areas by phase measurements alone because of the mutual interference of the radio waves propagating in different directions. The multi-path effect may have a different origin: refraction in the atmosphere, reflection from the earth's surface, diffraction or other effects. Characterizing this task as a wave problem allows determining dependence of the bending angle on the impact parameter. In this chapter the back-propagation method (BP), radio-holographic focused synthetic aperture approach (RHFSA), General Inversion Operator (GIO) transform, Fourier integral operators (FIO), spectral phase matching method (SPMM) will be considered. These methods use numerical techniques for determining the bending angle. It is difficult to compare the effectiveness of different methods. Therefore description of principal features of the aforementioned approaches will be examined in this chapter.

Radio holograms are the result of the registration of the amplitude and phase of the electromagnetic waves along the LEO trajectory during RO experiments. It is necessary to first obtain an altitude profile of the bending angle from analysis of the radio-hologram, and then to find dependence of the refractivity $N(h)$ on the height in the atmosphere and/or in the ionosphere.

Basis of the considered methods is presentation of the field along the LEO orbital trajectory as a superposition of the plane waves with the phase depending on the direction of propagation [68]. The electromagnetic field of a plane wave E may be written in a form

$$E(\omega,\mathbf{r},t) = A_0 \exp\left[i\left(-\omega\, t + \mathbf{kr} - \varphi\right)\right]. \quad (3.1.1)$$

where A_0 – is the wave amplitude, ω – is the circular frequency, $\mathbf{k}$ – is the wave vector determining the direction of wave propagation, $\mathbf{r}$ – is the spatial vector parallel to the direction from the center of the coordinate system ZOY - point O to the observation point, φ – is the phase of radio wave. Point O is disposed in the center of spherical symmetry of the atmosphere. Equation (3.1.1) may be presented in the form

$$E = A_w \exp\ (-\,i\ \omega\ t),\ A_w{=}A_0\exp[-\mathrm{i}(\mathbf{kr} - \varphi)\,], \quad (3.1.2)$$

$$\mathbf{kr} = \mathrm{k_Z z} + \mathrm{k_y y}, \quad (3.1.3)$$

$$\mathrm{k}^2 = 4\pi^2\lambda^{-2} = \omega^2\mathrm{c}^{-2} = \mathrm{k_Z}^2 + \mathrm{k_y}^2. \quad (3.1.4)$$

Eqs. (3.1.2) – (3.1.4) determine the frequency, direction of propagation, the amplitude and the phase of a plane wave. Equation (3.1.2) is a partial solution of the wave equation in free space

$$\frac{\partial^2 A_W}{\partial z^2}+\frac{\partial^2 A_W}{\partial y^2}+k^2 A_W=0\,. \tag{3.1.5}$$

After introducing the angle α between the wave vector $\mathbf{k}$ and direction OZ one can obtain

$$k_Z=k\ \cos\alpha\,,\quad k_y=k\ \sin\alpha\,. \tag{3.1.6}$$

It is convenient to introduce a new variable u

$$u=\sin\alpha\,. \tag{3.1.7}$$

After substitution of Eq. (3.1.7) in Eq. (3.1.6) next relationship can be obtained

$$k_Z=\pm\,k\left(1-u^2\right)^{1/2}\,. \tag{3.1.8}$$

The complex amplitude of the plane wave has a form

$$\begin{aligned} A_W(z,y)&=A_0\exp\ (i\varphi)\exp\left[\ i\,z\,k\left(1-u^2\right)^{1/2}\right]\ \exp(i\,u\,y),\ if\ z\geq 0\\ A_W(z,y)&=A_0\exp\ (i\varphi)\exp\left[-\,i\,z\,k\left(1-u^2\right)^{1/2}\right]\ \exp(i\,u\,y),\ if\ z<0 \end{aligned}\,. \tag{3.1.9}$$

The first equation (3.1.9) describes the plane waves with wave vector having a positive projection on the OZ axis. This equation corresponds to radio waves propagation during RO experiments. The second equation (3.1.9) is relevant to the plane waves propagating in the opposite directions. The field in the homogeneous part of medium may be presented as an integral sum of solutions of wave equation (3.1.9) [68]:

$$A_W(z,y)=\frac{1}{2\pi}\int_{-\infty}^{\infty}A(u)\ \exp\left[i\,k(z-z_0)\left(1-u^2\right)^{1/2}\right]\exp(i\,u\,y)\,du,\ z-z_0\geq 0 \tag{3.1.10}$$

Here $A(u)$ is a complex angular spectrum describing the amplitude and phase of the partial plane waves with the direction of propagation depending on the variable u. General solution (3.1.10) is corresponded to the initial condition on the straight line $z=z_0$, where the complex amplitude is equal to

$$A_W(z=z_0,y)=\frac{1}{2\pi}\int_{-\infty}^{\infty}A(u)\exp(i\,u\,y)\,du\,. \tag{3.1.11}$$

It is assumed that the straight line $z = z_0$ is located in free space and that the field inside the earth's ground is negligibly small. Expression (3.1.10) is the Fourier Integral Operator. By applying the Inverse Fourier transformation to equation (3.1.11), the function $A(u)$ is determined by using the known function $A_W(z, y)$ at $z = z_0$:

$$A(u) = \int_{-\infty}^{\infty} A_W(z = z_0, y)\ \exp(-i\ u\ y)\ dy\,. \tag{3.1.12}$$

The function $A(u)$ is the angular spectrum of the field. Relationship (3.1.10) indicates that the field in an arbitrary point of free space may be presented in the form of superposition of plane waves. Each plane wave has corresponding ray trajectory, a structure of the ray trajectories in any point of space is determined by the angular spectrum of the field.

The direct transform (3.1.10) will be used for analysis of RO data. Let us consider a 2-D problem with a scheme shown in Fig. 3.1.1. The point O corresponds to the earth's center and coincides with the center of the rectangular coordinate system ZOY. For simplicity, it will be assumed that the orbital planes of the satellite G, and satellite L coincide with the plane ZOY. This is not principal for the analysis of the RO experimental data and for understanding the problem of data analysis. Trajectory of the receiving satellite L during RO measurements is shown by curve L_1L_2. The amplitude and phase of the field are measured at the segment L_1L_2. If at point L only one ray intersects the segment L_1L_2, then the bending angle $\xi(p)$ between the direction of the ray wave vector **k** and OZ axis can be measured as a single-valued function of the impact parameter p. In some atmospheric conditions with special form of vertical profile of the refractivity $N(h)$, several rays may intersect the segment L_1L_2 in different directions, which corresponds to multi-path effect. Due to influence of the multi-path effect, measurements of the phase of RO signal do not allow finding of the bending angle as a single-valued function $\xi(p)$. Therefore, it is necessary to find location of the single-ray propagation areas where one should determine the amplitude and phase of RO signal. This area usually is disposed in a space between the real and the virtual focuses indicated in Fig. 3.1.1 by letters R and v, respectively. Determination of the field in a single-ray area is one of the main tasks of the back-propagation method.

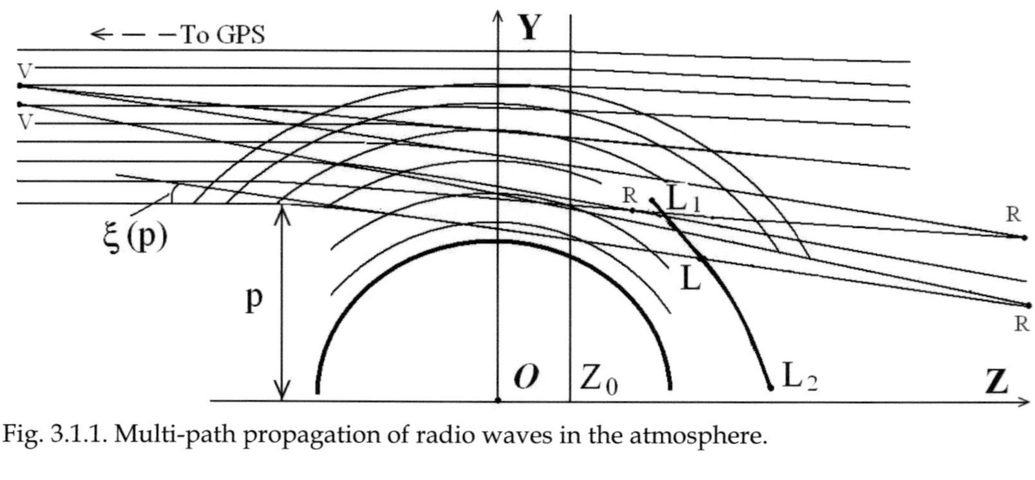

Fig. 3.1.1. Multi-path propagation of radio waves in the atmosphere.

3.2. Radioholographic focused synthetic aperture method (RHFSA)

The radio-holographic methods were designed with aim to extend an applicable domain of RO method and to increase its effectiveness, to exclude uncertainty in determing the bending angle due to multi-path propagation, and to heighten vertical resolution and accuracy in the retrieved atmospheric and ionospheric parameters [65, 69–81]. Below, the radioholographic focused synthetic aperture method (RHFSA) will be considered. RHFSA method uses the amplitude and phase of RO signal to obtain dependence of the bending angle $\xi(p)$ on the impact parameter p. It also uses appropriate information on the regular structure of the atmosphere and ionosphere for constructing a reference signal. For construction of reference signal, a model of the altitude dependence of the refractivity $N_m(h)$ is used [30, 31, 65, 69].

For consideration of RHFSA principle, let us introduce the angle γ between directions of ray and and orbital trajectory L_1L_2; for simplicity, we will consider only the case of circular orbit of the LEO and GPS satellites. The results of RO measurements are dependences of the phase path (eikonal), Doppler frequency, and amplitude over time, having significant variations due to atmospheric influence. Signal registered by LEO receiver is an envelope of the high-frequency field, which may be presented in the form

$$\mathrm{E}_r(t) = A(t)\ \exp\ [\ i\ k\ \psi(t)\ ], \tag{3.2.1}$$

where $k = \dfrac{2\,\pi}{\lambda}$ is the wave number, $A(t)$ is the amplitude, $\psi\ (t)$ is the eikonal.

In each point of the LEO satellite orbit the field $\mathrm{E}_r(t)$ is a sum of partial signals having the amplitude A_j, eikonal ψ_j, and arrival angle γ_j, each of them corresponds to an individual ray's trajectory

$$E_r(t) = \sum_{j=1}^{n} A_j(\gamma_j, t)\ \exp\ \left[i\ k\ \psi_j(\gamma_j, t)\ \right], \tag{3.2.2}$$

where n is the number of rays' trajectories, connecting the points G and L, ψ_j is the eikonal for j –th ray, determined by relationships:

$$\begin{aligned} &\psi_j(p_j) = L(p_j) + \kappa(p_j), \qquad L(p_j) = d_{1j} + d_{2j} + p_j\ \xi(p_j) \\ &\xi(p_j) = -d\kappa(p_j)/dp_j \\ &d_{1j} = (r_l^2 - p_j^2)^{1/2}, \qquad d_{2j} = (r_g^2 - p_j^2)^{1/2} \end{aligned} \tag{3.2.3}$$

where $p_j(t)$, $\xi(p_j)$, $\kappa(p_j)$ are the impact parameter, bending angle, and main refractive part of the phase path, corresponding to j –th ray, $L(p_j)$ is the length of geometric part of the phase path, consisting of the two length d_{1j}, d_{2j} and the arc $p_j\xi(p_j)$. Main refractive part $\kappa(p_j)$ accounts for a change in the ray's path due to influence of vertical gradient of refractivity.

The left side of the Eqs. (3.2.1) and (3.2.2) should be equal, therefore an equation may be obtained that establishes a relationship between measured parameters of RO signal $A(t)$ and $\psi(t)$ and unknown values $A_j(\gamma_j,t)$ and $\psi(\gamma_j,t)$, thus characterizing the angular spectrum of the field as

$$A(t)\exp\left[\mathrm{i}k\psi(t)\right] = \sum_{j=1}^{n} A_j(\gamma_j,t)\exp\left[\mathrm{i}k\psi_j(\gamma_j,t)\right]. \tag{3.2.4}$$

Aim of ana;ysis of RO data is to find the arrival angles γ_j and the bending angles ξ_j. For solution of equation (3.2.4) it is necessary to apply a radio-holographic principle and to form a reference signal having the amplitude equal to unity

$$E_m(t) = \exp\left[i\,k\psi_m(t)\right], \tag{3.2.5}$$

where ψ_m is the eikonal, corresponding to the altitude profile $N_m(h)$, ψ_m is evaluated for real orbits of the satellites

$$\psi_m(t) = \int_{G}^{L_j} [1+N_m(h)\cdot 10^{-6}]\,dl. \tag{3.2.6}$$

The eikonal $\psi_m(t)$ may be determined after substitution in (3.2.3) dependences of the impact parameters $p_m(t)$ and the refractive angle $\xi_m(t)$ on time, found by use of the profile $N_m(h)$ and known values of the satellites velocities and coordinates. Integration in (3.2.6) is fulfilled along the ray GL_j, h is the altitude relative to the earth's surface, dl is an element of the ray trajectory. The ratio of the registered and reference signals according to (3.2.4) and (3.2.5) is equal to

$$\begin{aligned}&\mathrm{E}(t) = \mathrm{E}_r(t)\mathrm{E}_m^{-1}(t) = \\ &A(t)\,\exp\,\left[i\,k\,(\psi(t)-\psi_m(t))\right] = \sum_{j-1}^{n} A_j(\gamma_j,t)\exp\left[i\,k\,(\psi_j(\gamma_j,t)-\psi_m(t)\right]\end{aligned} \tag{3.2.7}$$

By multiplication of the right and left part of equation (3.2.7) on $\exp(-i\omega t)$ and after integration over time in the interval $-\Delta t/2 \le t \le \Delta t/2$, one can obtain spectrum of the ratio of measured and reference signals:

$$W(\omega) = \sum_{j=1}^{n} \int_{-\Delta t/2}^{\Delta t/2} A_j(\gamma_j)\exp\{i\,[k(\psi_j(\gamma_j,t)-\psi_m(t))-\omega\, t]\}\,dt, \tag{3.2.8}$$

where $\omega = 2\pi f$ – is the spectral frequency. Let us assume that only one ray will correspond to each frequency in the spectrum $W(\omega)$. Multi-path in most cases is originating owing to the influence of the layers structures in the lower atmosphere, so the phase paths of rays in the middle and upper atmosphere are nearly coinciding. This allows subtracting by use of a reference signal influence of the main part of the phase paths, common for all rays, thus significantly compressing the spatial spectrum of radio waves. The reference signal is essentially a matching heterodyne applied to measurements of deflection of the impact parameter and bending angle of the main signal from corresponding parameters of the reference signal, and for detecting the weak signals. This method is analogous to the focused synthetic aperture technique in radio location, which is applied for separation of the signals from targets and determining their coordinates by use of matching filtration. As a result of compressing the spatial spectrum, one may observe reflected from the earth surface weak signals, having distinction only in the Doppler frequency as compared with the main intensive signal. Under optimal form of the reference signal, one may observe in the spectrum $W(\omega)$ a sharp peak that corresponds to the compressed spectrum of the main signal and weak signals being notable because of the Doppler shifts. Difference of the phase in the right part of (3.2.7) may be presented as the Taylor series centered at $t=0$, and containing in the interval $-\Delta t/2 \le t \le \Delta t/2$ only the terms, linearly depending on time:

$$k\,\frac{d}{dt}\left[\psi_j(\gamma_j,t)\;-\;\psi_m(t)\right]=\omega_j, k\,[\psi_j\;(\gamma_j,t)\;-\;\psi_m(t)\;]=\omega_{j0}t+\phi_{j0}\,, \qquad (3.2.9)$$

where ω_{j0} , φ_{j0} are the constants in the interval of integration Δt . If changes in the amplitude $A_j(\gamma_j)$ of partial rays are small then integration in equation (3.2.8) may be fulfilled analytically with accounting for linear dependence on time of the eikonal differences in (3.2.9). After integration one may obtain from (3.2.8)

$$W(\omega)=\sum_{j-1}^{n} A_j\;(\gamma_j)\;\Delta t\;F\;(\omega,\omega_j)\,, \qquad (3.2.10)$$

$$F(\omega,\omega_j)=\left[\frac{\Delta t\;(\omega-\omega_j)}{2}\right]^{-1}\sin\left[\frac{\Delta t\;(\omega-\omega_j)}{2}\right]\;\exp\;(i\;\varphi_{j0})\,, \qquad (3.2.11)$$

where ω_j is the frequency, corresponding to j-th partial ray. According to (3.2.11), the function $F(\omega,\omega_j)$ forms in spectrum $W(\omega)$ main maximum, and some weak maxima at frequencies ω_j . Function $F(\omega,\omega_j)$ achieves a maximum at $\omega=\omega_j$ and then quickly diminishes when the absolute value of difference $\omega=\omega_j$ increases. The width of maximum on the one half power level which determines vertical resolution is nearly equal to $2/\Delta t$. When interval of integration is extended the resolution increases. Value Δt is bounded by an assumption of small variations in the amplitude, initial phases and frequencies of partial signals inside the integration interval. Therefore, if the maxima of partial signals in spectrum $W(\omega)$ do not overlap, the described technique gives a solution of problem of detecting partial rays in the case of interfering influence of the main signal. The Doppler frequency ω may be recalculated to the arrival angles γ observed at the time instant $t = 0$ by use of relationship

$$\gamma = \sin^{-1}\left[(\omega - \omega_m)(kv)^{-1} + \sin\gamma_m\right], \tag{3.2.12}$$

where $v = r_l \dfrac{d\theta}{dt}$, r_l is the distance OL2, ω_m, γ_m are the Doppler frequency and arrival angle, corresponding to the reference signal. With accounting for the relationship between the angular frequency and arrival angle (3.2.12), the absolute value of spectrum $W(\omega)$ describes an angular power distribution of radio waves, registered at the time instant $t = 0$. The angular power distribution corresponds to 1-D distributions of radio brightness (or 1-D radio image of vertical structure of the atmosphere) along the line TO, observed from the side of LEO satellite :

$$A_1(\gamma_j(\omega)) = \left| W^2(\omega)\ (\Delta t)^{-2} \right| \tag{3.2.13}$$

where $A_1(\gamma_j)$ is the angular power spectrum of the wave field, the right part (3.2.13) describes the frequency distribution of the received signal energy. On the basis of this algorithm a radio-holographic method was elaborated and validated for obtaining radio image of the atmosphere and signals, reflected from the earth's surface [65, 71-74]. This method may be applied to obtain vertical profiles of the atmospheric parameters from measurements of dependence of the bending angle $\xi(p)$ on the impact parameter p [70]. To obtain this dependence it is necessary to measure the Doppler frequency ω_0 of the maximum in spectrum of the main atmospheric signal. From the Doppler shift ω_0 one may determine a value of the impact parameter $p = p_0$, corresponding to the time instant $t = 0$, i.e. to the middle of the integration interval [65,69,70]

$$p = p_m - f_0\lambda\left[\frac{d\theta}{dt}\right]^{-1}, \quad f_0 = \frac{\omega_0}{2\pi} \tag{3.2.14}$$

where λ is the wavelength in free space, p_m – is the impact parameter of the reference signal. Then one may find the bending angle as a function of the impact parameter p

$$\xi(p) = \xi_m(p_m) + \sin^{-1}(pr_g^{-1}) + \sin^{-1}(pr_l^{-1}) - \sin^{-1}(p_m r_g^{-1}) - \sin^{-1}(p_m r_l^{-1}), \tag{3.2.15}$$

where ξ_m is the bending angle, corresponding to the altitude profile $N_m(h)$.

Equation (3.2.13) determines the energy of components in the angular spectrum, which are coherent with the reference signal. Maximum of the angular spectrum corresponds to location of the «main» ray and in accordance with equation (3.2.14) determines its impact parameter p and bending angle ξ. The width of angular spectrum on the half power level characterizes the root mean square error in determining the bending angle and corresponding atmospheric parameters. After determining dependence $\xi(p)$ a standard Abel transformation method is used to find vertical profiles of the refractivity N and temperature T. Radio-holographic principle improve vertical resolution as compared with the Doppler shift method. According to (3.2.11), the angular uncertainty $\Delta\gamma$ and corresponding vertical resolution Δh depend on the time of coherent analysis of RO signal Δt. According to (3.2.10) and (3.2.14) one has

$$\Delta\gamma = \lambda\ (2\ v\ \Delta t\ \cos\gamma_m)^{-1}\ ,\qquad \Delta h = \Delta p = \lambda\ r_1 (2\ v\ \Delta t)^{-1}. \tag{3.2.16}$$

According to (3.2.16), the angular uncertainty $\Delta\gamma$ and corresponding vertical resolution Δh of the radio-holographic method are proportional to the wavelength λ in distinction from the Fresnel resolution, which is proportional to the square root of the wavelength λ.
The form of spectra (3.2.10) and (3.2.13) gives a demonstration of conditions of radio waves propagation through the atmosphere in a single-ray area. In a single-ray area the spectra $W(\omega)$ and $A_1(\gamma)$ have one maximum; in a multi-path area they have several maxima or a broad spectrum. The angular spectrum in RHFSA method may be interpreted as a «radio-image» or a «distribution of radio brightness» in the atmosphere and ionosphere, observed from the orbit of the satellite – receiver. The next features characterize spectra $W(\omega)$: at the altitudes H = 40…15 km one sharp spectral line is observed. During immersion of ray in the troposphere additional spectral components may appear. These components are caused by reflection of radio waves from the earth's surface and multi-path propagation. Radio-holographic method has high vertical resolution Δh, therefore, the altitude profiles of the temperature $T(h)$ and electron density $N_e(h)$ found by this method have more details. Radio-holographic method was applied for determining the altitude profiles of the refractivity and temperature [70]. The application of the radio-holographic approach to the analysis of RO data, obtained by *MICROLAB* satellite, allows detection and observation of reflections from earth's surface and multi-path propagation in the troposphere [71–74].
It is difficult to derive the reference signal in the lower troposphere, which may compensate the temporal change of the Doppler shifts for all partial rays. For application of the radio-holographic method in this situation it is necessary to modernize it. This modernization may be conducted on the basis of the Full Spectral Inversion (FSI) method proposed in [75] and modernized in [76]. Temporal dependence of the eikonal corresponding to the reference signal may not fully follow the changes of the eikonals relevant to the partial signals. As a consequence a compensation of the contributions of different parts of the integration interval in the spectrum takes place, and the stationary phase points are more important for evaluating the spectrum $W(\omega)$ in (3.2.8). Stationary phase point is determined by equation:

$$\omega = k\ \frac{d}{dt}\ [\psi_j(\gamma_j, t) - \psi_m(t)] \tag{3.2.17}$$

Equation (3.2.17) gives in an implicit form the time instant $t(\omega)$ as a function of the Doppler shift ω . It is supposed that dependence $t(\omega)$ is single-valued, i.e. each value of the frequency ω may appear at one instant of time in the integration interval Δt. In the case of multi-path situation, the inverse function $\omega(t)$ may be multivalued since at one instant of time several rays intersect at different angles an orbit of the satellite L in a given point. For each stationary point in the integration interval a set of Eqs. can be obtained from (3.2.8):

$$W(\omega) = B_j(\gamma_j, t)\ \exp\ [i\Phi\ (t(\omega)\ )\ ],\ \Phi(t(\omega)) = k\ [\psi\ (\gamma_j, t) - \psi_m(\gamma_m, t) - \omega t \quad (3.2.18)$$

$$B_j(\gamma_j, t) = A_j(\gamma_j, t)\ S(t); \qquad S(t) = (2\pi)^{1/2} \left\{ -k \frac{d^2}{dt^2} \left[\psi(\gamma_j, t) - \psi_m(\gamma_m, t) \right] \right\}^{-1/2},$$

where $B_j(\gamma_j, t)$, $\Phi\ (t(\omega)\ )$ are the amplitude and phase of the spectrum components $W(\omega)$ corresponding to the stationary point, $S(t)$ is a factor accounting for contribution of the stationary phase point. Because the function $t(\omega)$ is single-valued in the interval Δt, the sum (3.2.8) gives only one term corresponding to a partial ray, intersecting the satellite trajectory and having in a given instant of time t the Doppler shift ω. The time instant t may be estimated from the derivative of the phase of spectrum $\Phi(t\ (\omega)\ )$ with respect to frequency ω

$$\frac{d\Phi\ (t\ (\omega)\ )}{d\omega} = \left[k\ \frac{d}{dt}\ (\psi_j(\gamma_j, t) - \psi_m(t)) - \omega \right] \frac{dt(\omega)}{d\omega} - t = -t\ . \qquad (3.2.19)$$

The square bracket on the right-hand side of (3.2.19) is equal to zero because of fulfilling the equation (3.2.17) in a stationary phase point. One may determine by use of (3.2.14) the impact parameter p, corresponding to the time instant $t = -d\Phi(t(\omega))/d\omega$. Then the bending angle $\xi(p)$ may be found from relationship (3.2.15) as a function of the impact parameter p. Vertical resolution is determined in this case by the difference of the second derivatives of the phases of the reference and registered signals with respect to time. If the second derivatives coincide, then vertical resolution is near the maximal value as determined from relationship (3.2.16).

In papers [76, 77], a spectral phase matching method (SPMM) has been proposed to determine directly dependence of the bending angle on the impact parameter. The SPMM technique is an important case of the radio-holographic method. The SPMM technique may be applied for general case of non-circular orbits transmitting and receiving satellites. The SPMM technique uses a special form of the reference signal:

$$E_0(t) = \exp\ \left[i\ k\ L_0(t) \right], \qquad (3.2.20)$$

where L_0 is the eikonal value, corresponding to the geometric part of the phase path under fixed value of the impact parameter p_0

$$L_0(p_0)\ =\ \sqrt{R_l^{\,2} - p_0^{\,2}} + \sqrt{R_g^{\,2} - p_0^{\,2}} + p_0 \xi(p_0)\,. \qquad (3.2.21)$$

The bending angle $\xi(p_0)$ in (3.2.21) may be determined on known values of the central angle θ and parameters R_g, R_l

$$\xi(p_0) = \theta - \cos^{-1}(p_0 / R_l) - \cos^{-1}(p_0 / R_g). \qquad (3.2.22)$$

Parameters R_g, R_l, θ in (3.2.21), (3.2.22) depend on time. Derivative of $L_o(p_o)$ with respect to fixed parameter p_0, may be found from Eqs. (3.2.21), (3.2.22)

$$dL_0(p_0) / dp_0 = \xi(p_0) \qquad (3.2.23)$$

because θ in (3.2.22) does not depend on p_0. The impact parameter p_0, in the general case, does not coincide with the impact parameter p corresponding to ray GTL. The next relationships are fulfilled for ray GTL

$$\begin{gathered}\psi(p) = L(p) + \kappa(p),\ L(p) = \sqrt{R_l^2 - p^2} + \sqrt{R_g^2 - p^2} + p\xi(p), \\ \xi(p) = \theta - \cos^{-1}(p / R_l) - \cos^{-1}(p / R_g),\ \xi(p) = -d\kappa(p) / dp\end{gathered} \qquad (3.2.24)$$

where $\kappa(p)$ is the main refractive part of the phase path. The Doppler frequencies $f(p), f_0(p_0)$ corresponding to the ray GTL and reference signal, may be determined from the relationships [65,76]

$$\begin{gathered}f(p) = d\psi / dt = \\ R_l^{-1}\sqrt{R_l^2 - p^2}dR_l / dt + R_g^{-1}\sqrt{R_g^2 - p^2}dR_g / dt + pd\theta / dt \\ f_0(p_0) = dL_0 / dt = \\ R_l^{-1}\sqrt{R_l^2 - p_0^2}dR_l / dt + R_g^{-1}\sqrt{R_g^2 - p_0^2}dR_g / dt + p_0 d\theta / dt\end{gathered} \qquad (3.2.25)$$

Ratio of the registered and reference signals may be presented in the form:

$$E(t) = A(t)\ \exp\left[i\, k\left(\psi(t) - L_0(t)\right)\right]. \qquad (3.2.26)$$

Integration of (3.2.26) on time in the interval ΔT of RO measurements $0 \le t \le \Delta T$

$$W(p_0) = \int_0^{\Delta T} A(t)\ \exp\ [i\, k\, \Phi(t)]dt,\ \Phi(t) = \psi(t) - L_0(t) \qquad (3.2.27)$$

gives the spectrum $W(p_0)$ depending on the impact parameter p_0. Let us assume that for the time of integration ΔT each partial signal has own value of the impact parameter p_j, which does not coincide with the impact parameters of other signals. Then only one ray relates to each value p_0 in a spectrum $W(p_0)$. Value p_0 corresponding to a stationary point $t(p_0)$ of the integral (3.2.27) introduces the main contributions in spectrum $W(p_0)$. It is supposed that the function $t(p_0)$ is single-valued. The stationary phase point equation is

$$d[\psi(t)-L_0(t)]/dt=0\,. \tag{3.2.28}$$

Equation (3.2.28) determines the time instant $t(p_0)$ in an implicit form corresponding to a stationary phase point. By use of relationship (3.2.25), equation (3.2.28) may be written in the form:

$$\begin{aligned}&\{-(p+p_0)[R_l^{-1}dR_l/dt(\sqrt{R_l^2-p^2}+\sqrt{R_l^2-p_0^2})^{-1}\\&+R_g^{-1}dR_g/dt(\sqrt{R_g^2-p^2}+\sqrt{R_g^2-p_0^2})^{-1}]+d\theta/dt\}(p-p_0)=0\end{aligned} \tag{3.2.29}$$

It follows from Eqs. (3.2.28) and (3.2.29), that value p is equal to p_0 in the stationary phase point:

$$p=p_0 \tag{3.2.30}$$

According to (3.2.30) value p_0 is the impact parameter at the time instant t.

Spectrum $W(p_0)$ in the stationary phase point of the integral (3.2.27) has the form:

$$\begin{aligned}&W(p_0)=\sqrt{2i\pi k^{-1}}A(p_0)/(d^2\Phi/dt^2)^{1/2}\exp[ik\Phi(p_0)]\\&\Phi(p_0)=\psi(p_0)-L_0(p_0)\end{aligned} \tag{3.2.31}$$

Derivative of the phase of spectrum $W(p_0)$ with respect to parameter p_0 gives the bending angle

$$d\Phi(p_0)/dp_0=\{d[\psi(p_0)-L_0(p_0)]/dt\}dt/dp_0-\partial L_0(p_0)/\partial p_0=-\xi(p_0)=-\xi(p)\,. \tag{3.2.32}$$

Relationship (3.2.32) follows from equation (3.2.29) for the stationary phase point of integral (3.2.27). Therefore, the phase matching method allows finding dependence of the bending angle on the impact parameter by use of differentiation of the phase of spectrum $W(p_0)$ with respect to parameter p_0.

A feature of the phase matching method is possibility to determine the total absorption of radio waves from measured spectrum $W(p_0)$ [77]. According to [53] the power of RO signal is proportional to the refractive attenuation $X(p)$, which may be found from relationship:

$$X(p)=pR_0\left[p_s\sqrt{R_g^{\,2}-p^2}\sqrt{R_l^2-p^2}\left|\frac{\partial\theta}{\partial p}\right|\right]^{-1}, \tag{3.2.33}$$

where R_0 is the distance GL, p_s is the impact parameter corresponding to the line of sight GL. From the other side, value $\frac{\partial\theta}{\partial p}$ may be determined from a ratio [22]:

$$\frac{\partial\theta}{\partial p} = B(t)(dp/dt)^{-1},$$
$$B(t) = d\theta/dt - p(R_l\sqrt{R_l^2 - p^2})^{-1}dR_l/dt \; - p(R_g\sqrt{R_g{}^2 - p^2})^{-1}dR_g/dt \tag{3.2.34}$$

Contribution of the stationary point depends on the second derivative of the phase of the function $\Phi(t)$ with respect to time in the integral (3.2.27). With accounting for (3.2.29) one may find:

$$d^2\Phi/dt^2 = B(t)dp/dt \tag{3.2.35}$$

By use of relationships (3.2.33)-(3.2.35) one may find the ratio $A(p_0)/(d^2\Phi/dt^2)^{1/2}$ in equation (3.2.31):

$$A(p_0)/(d^2\Phi/dt^2)^{1/2} = A_0(X/d^2\Phi/dt^2)^{1/2}$$
$$= A_0[B(t)]^{-1}(pR_0)^{1/2}(p_s\sqrt{R_g{}^2 - p^2}\sqrt{R_l^2 - p^2})^{-1/2} \approx const \tag{3.2.36}$$

where A_0, p_s is the radio wave amplitude and the impact parameter in the free space, respectively. dependence of ratio $A(p_0)/(d^2\Phi/dt^2)^{1/2}$ on the impact parameter p is weak in RO experiments. Therefore the refractive attenuation is practically excluded from spectrum $W(p_0)$. The phase matching method may be used for measurements of the total absorption of radio waves [76, 77].

3.3. GIO, CT and BP methods

Application of the integral operators to analysis of RO data results from a necessity to remove influence of multi-path distortion on the RO inversion. In some cases this can be achieved by using the special radio-holographic methods. In this section an application of the Fourier Integral Operators (FIO) based on the Zverevs' integral transformations [68] is considered for analysis of RO data. This consideration will be provided by use of results obtained in [74] for a geometrical scheme shown in Fig. 3.3.1. Let us introduce the rectangular coordinate system ZOY with center at point O, which coincides with the center of spherical symmetry of the atmosphere. A LEO satellite and emitting satellite are located at point P and G, respectively. Axis OZ coincides with direction GO. Orbital trajectory of LEO satellite is indicated by curve PP'. Trajectory of radio waves is shown by curve GTP, where T is the radio ray perigee. Radio waves can propagate in different directions along rays which may intersect at given point P in the case of multi-path situation . Multi-path effect makes difficult application of the geometrical optics method to determine dependence of the field phase on the impact parameter and introduces distortions to the retrieved altitude dependence of the bending angle.

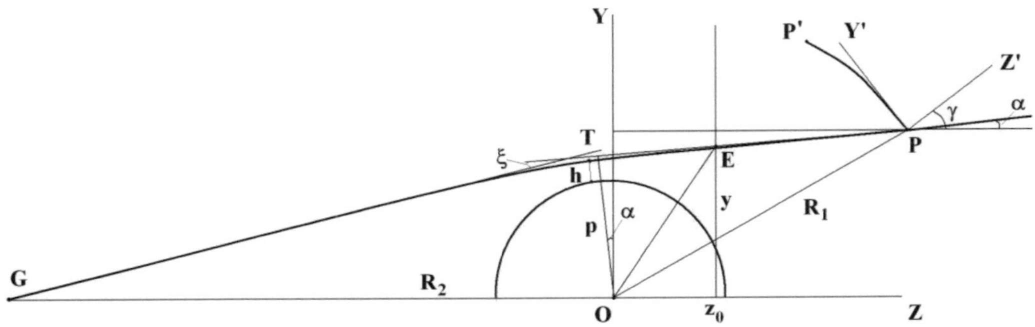

Fig. 3.3.1. Geometrical parameters for GIO, RFSA, CT and BP methods.

For description of the electromagnetic field near the ray trajectory P'P we apply the Zverev transform [68], which connects the field $E(y,z)$ and its angular spectrum in a homogeneous part of medium (in free space):

$$E(y,z) = \int du\, A(u)\exp\{ik[\Phi(u,y,z) - \Phi_j(u)]\}, \tag{3.3.1}$$

$$\Phi(u,y,z) = z\sqrt{1-u^2} + yu \ , \tag{3.3.2}$$

$$u = \sin\alpha,\ \sqrt{1-u^2} = \cos\alpha \ , \tag{3.3.3}$$

where $\Phi(u,y,z), \Phi_j(u)$ are the phase function and initial phase of the *j-th* physical ray, u is the projection of the unit wave vector of a plane wave on the direction OY, α is the angle between the direction of plane wave propagation and axis OZ.

With aim to remove influence of multi-path propagation on the retrieved dependence of the bending angle on the impact parameter let us transform the field $E(y,z)$ from coordinate y to coordinate p. By using the transformed field $E(p,z)$, one can obtain the single-valued dependence of the bending angle on the impact parameter. To connect the Zverev's transform with the LEO trajectory let us introduce the new co-ordinate system y', z' with the centre at point *P* and oriented at the angle γ relative to the co-ordinate system *y, z* (Fig. 3.3.1). The coordinate y' is reckoned from point *P* along the tangent to orbital trajectory of LEO satellite:

$$y = y_P + y'\cos\gamma + z'\sin\gamma;\ z = z_P + z'\cos\gamma - y'\sin\gamma, \tag{3.3.4}$$

where y_p, z_p are the co-ordinates of point *P* (Fig. 3.4) in co-ordinate system *ZOY* (Fig. 3.3.1). We introduce a General Inversion Operator (GIO) transform [74] as a Fourier Integral Operator (FIO) $I(p)$:

$$E(p) = I(p)E(y,z) = \int d\eta B(\eta)\exp\{i\kappa[pf(\eta) - d(\eta)]\} I_1(\eta), \tag{3.3.5}$$

$$I_1(\eta) = \frac{k}{2\pi}\int ds \exp(-ik\eta s) E[y(s), z(s)], \tag{3.3.6}$$

where $I_1(\eta)$ is the internal Fourier-operator, $B(\eta)$ is the amplitude function, $f(\eta)$, $d(\eta)$ are the auxiliary functions, p is the parameter, having different physical interpretation depending on the form of $f(\eta)$, s is the path of integration along the orbital trajectory of the LEO satellite. Below we assume that a LEO orbital trajectory during RO measurements can be approximated by a straight line and the integration on s can be changed by integration on the variable y' in the operator $I_1(\eta)$ (3.3.6). Practically this assumption is not significant because during the time interval of RO measurements the orbital trajectory of LEO satellite can be considered as nearly coinciding with a segment of a straight line. GIO transform can be considered as a generalization of the FIO's introduced in [78, 79]. After substitution (3.3.1), (3.3.4), into (3.3.6) one can obtain letting $z'=0$:

$$I_1(\eta)=\frac{k}{2\pi}\int dy'\exp(-ik\eta y')\int du\, A(u)\exp\{ik[y'\sin(\alpha-\gamma)+z_p\cos\alpha+y_p\sin\alpha-\Phi_j(u)]\} \quad (3.3.7)$$

After changing of integration order on y' and u in (3.3.7) the next equation is followed:

$$\delta[\eta-\sin(\alpha-\gamma)]=\frac{k}{2\pi}\int dy'\exp\{iky'[\sin(\alpha-\gamma)-\eta]\}, \quad (3.3.8)$$

$$I_1(\eta)=\int du\, A(u)\delta[\eta-\sin(\alpha-\gamma)]\exp\{ik[z_p\sqrt{1-u^2}+y_p u-\Phi_j(u)]\}, \quad (3.3.9)$$

where $\delta[\eta-\sin(\alpha-\gamma)]$ is the delta-function. Substitution of (3.3.9) into (3.3.5) gives after integration on η the next result:

$$E(p)=\int du\, A(u)B(\eta)\exp\{ik[pf(\eta)-d(\eta)+z_p\cos\alpha+y_p\sin\alpha-\Phi_j(u)]\} \\ \eta=\sin(\alpha-\gamma) \quad . \quad (3.3.10)$$

The left part of (3.3.10) is the field $E(p)$ transformed by the operator $I(p)$ from RO signal. Function $d(\eta)$ in (3.3.10) is arbitrary, and one can chose $d(\eta)$ to simplify the GIO transform.

$$d(\eta)=(z_p-z_0)\cos(\gamma+\sin^{-1}\eta)+y_p\sin(\gamma+\sin^{-1}\eta), \quad (3.3.11)$$

where z_0 is a parameter which determines the location of a straight line where the field should be retrieved. If the origin of the co-ordinate system y', z' is disposed at the OZ axis and $\gamma=0$, $y_p=0$ then the function $d(\eta)$ is equal to $d(\eta)=(z_p-z_0)(1-\eta^2)^{1/2}$ and coincides with the phase of the transfer function for free space introduced earlier [68]. Substitution of (3.3.11) into (3.3.10) gives

$$E(p,z_0)=\int du A(u)B(\eta)\exp\{i\kappa[pf(\eta)+z_0\cos\alpha-\Phi_j(u)]\},\ \alpha=\sin^{-1}u \quad . \quad (3.3.12)$$

The relationship (3.3.12) describes the result of transformation of the field $E(y,z)$ by the GIO operator $I(p)$. Below we will consider two important results: the first corresponds to the case of back-propagation (BP) method, the second is relevant to the canonical transformation (CT). The BP case can be obtained from (3.3.5) and (3.3.12) by choosing $B(\eta)$ and $f(\eta)$ in the form

$$f(\eta) = \sin(\gamma + \sin^{-1}\eta),\ B(\eta) = 1. \tag{3.3.13}$$

After substitution of (3.3.13) into (3.3.12) one can obtain

$$\mathrm{E}(p, z_0) = \int \mathrm{d}u A(u) \exp\{i\kappa[\mathrm{p}u + z_0\sqrt{1-u^2} - \Phi_j(\mathrm{u})]\}\ . \tag{3.3.14}$$

In this case the right part in (3.3.14) describes the field distribution along the straight line $z = z_0$ and coincides with (3.3.1). As a consequence, p has a geometrical sense of coordinate y (Fig. 3.3.1). The Fourier Integral Operator (3.3.5), (3.3.6) restores the field along the straight line $z = z_0$ by use of transformation of the field measured along the tangent PY' to orbital trajectory at point P. It is assumed that the straight line $z = z_0$ is located in an area of one ray propagation. Dependence of the bending angle on the impact parameter may be found by using the stationary phase principle. By utilizing this principle the field $E(p, z_0)$ can be presented in the form

$$E(y, z_0) = \sqrt{\frac{-2i\pi}{kr_s}} A(u_s) \cos\alpha_S \exp[ik\Phi_s(y,\alpha_s)],\ u_s = \sin\alpha_s\ , \tag{3.3.15}$$

$$r_s = z_0 \cos\alpha + y \sin\alpha + d^2\Phi_j / d\alpha^2,\ \alpha = \alpha_s, \tag{3.3.16}$$

where r_s is the curvature radius of the wave front at the point with coordinates y, z_0, α_s is connected with coordinates y, z_0 by the stationary phase equation

$$y = \{z_0\ tg\alpha + [d\Phi_j(\sin\alpha)/d\alpha]/\cos\alpha\}\ (\alpha = \alpha_s). \tag{3.3.17}$$

The function $\Phi_s(y,\alpha_s)$ is the phase of the field $E(p, z_0)$ and may be estimated from relationship

$$\Phi_s(y,\alpha_s) = z_0\cos\alpha_S + y\sin\alpha_s - \Phi_j(\sin\alpha_s). \tag{3.3.18}$$

Differentiation of the phase $\Phi_s(y,\alpha_s)$ (3.3.18) with respect to the coordinate y gives a next relationship between the angle α_s and y in the stationary phase point of the integral (3.3.14)

$$\alpha_s = \sin^{-1}[d\Phi_s(y,\alpha_s)/dy]. \tag{3.3.19}$$

The angle α_s determines a slope of RO ray relative to the OZ axis in the point with coordinates y, z_0. From known values of the angle α_s and coordinates y, z_0 the impact parameter p may be estimated as

$$p = y\cos\alpha_s - z_0 \sin\alpha_s \tag{3.3.20}$$

and then the bending angle $\xi(p)$ may be found if the radius of circular orbit of transmitting GPS satellite r_g is known

$$\xi(p) = \sin^{-1}(p/r_g) - \alpha_s(p). \tag{3.3.21}$$

When the multi-path effect is absent along the straight line $z = z_0$, vertical resolution is increasing because the distance of the point $z = z_0$ from the ray perigee T is essentially smaller than PT (Fig. 3.3.1). Therefore the GIO corresponding to the back-propagation method (3.3.5), (3.3.6) has a special form

$$\begin{aligned} E(y,z_0) = I_b(y)E(y,z) = \int d\eta \exp\{i\kappa[(y-y_p)\sin(\gamma+\sin^{-1}\eta) \\ +(z_0-z_p)\cos(\gamma+\sin^{-1}\eta)]\}I_1(\eta) \end{aligned} \tag{3.3.22}$$

$$I_1(\eta) = \frac{k}{2\pi}\int ds \exp(-ik\eta s)E[y(s), z(s)]. \tag{3.3.23}$$

The integration limits on the variable s are bounded by the length of orbit segment PP′ which corresponds to the interval of measurements of the RO amplitude and phase. The limits of integration on the variable η are bounded by an inequality $-1 \le \eta \le 1$.
Let us compare Eqs. (3.3.22), (3.3.23) with known back-propagation equation [80]

$$E(y,z_0) = \sqrt{\frac{k}{2\pi}}\int ds\, D^{-1/2}\cos\varphi \exp(i\pi/4 - ikD)E(s), \tag{3.3.24}$$

where D is the distance between the current integration point s on an orbital trajectory of LEO satellite and observation point having coordinates y, z_0; φ is the angle between normal to the curve $P'P$ in point s and direction to the observation point. Note the distinction between the formulas (3.3.22), (3.3.23) and (3.3.24) for BP field. This is connected with difference in the type of own functions used for derivations (3.3.22), (3.3.23) and (3.3.24) (plane and spherical waves in free space, respectively). The formula (3.3.24) is valid for any curve in free space, Eqs. (3.3.22), (3.3.23) describe the field distribution along the straight lines. Both formulas restore the field along the straight rays intersecting the orbital trajectory. Inside the medium the restored field may not coincide with real field because the used Green functions are corresponding to free space. Coincidence with the real fields can take place only on the orbital trajectory of LEO satellite [78, 79, 93]. However the directions of the tangents to the rays and values of the bending angles coincide for the real and restored fields.

In the second important partial case of the GIO transform the functions $f(\eta)$ and $d(\eta)$ can be chosen in the form

$$f(\eta) = \gamma + \sin^{-1}\eta,\ d(\eta) = z_p \cos(\gamma + \sin^{-1}\eta) + y_p \sin(\gamma + \sin^{-1}\eta), \tag{3.3.25}$$

In this case the GIO transform may be described by Eqs.

$$E(p,\gamma) = I_c(y)E(y,z) = \int d\eta B(\eta) \exp\{i\kappa[p(\gamma + \sin^{-1}\eta) - z_p \cos(\gamma + \sin^{-1}\eta) \\ -y_p \sin(\gamma + \sin^{-1}\eta)]\} \mathrm{I}_1(\eta) \tag{3.3.26}$$

where $I_1(\eta)$ is determined by relation (3.3.23). Relationship (3.3.26) may be rewritten in the form

$$E(p,\gamma) = I_c(y)E(y,z) = \\ \exp(ikp\gamma)\int d\eta B(\eta)\exp[i\kappa(p\sin^{-1}\eta - z_{p\gamma}\sqrt{1-\eta^2} - y_{p\gamma}\eta)]\, \mathrm{I}_1(\eta) \tag{3.3.27}$$

$$y_{p\gamma} = y_p \cos\gamma - z_p \sin\gamma = 0;\ z_{p\gamma} = z_p \cos\gamma + y_p \sin\gamma = r_{l0}, \tag{3.3.28}$$

where $y_{p\gamma}$, $z_{p\gamma}$ are the coordinates of the point P in the coordinate system rotated by the angle γ relative to the ZOY system and with the center at point O, r_{l0} is the distance OP. For the case $y_1 = 0,\ \gamma = 0,\ B(\eta) = (1-\eta^2)^{-1/4}$ the formula (3.3.27) coincides with the CT transformation found earlier [81]. The CT method [81] utilizes the next assumptions: (1) the orbital segment where the field is measured is perpendicular to the direction GO, (2) the point P is located on the straight line GO, and (3) the angle γ is equal to zero. In real conditions the RO orbit is significantly inclined relative to the straight line GO and $\gamma \neq 0$. In the case of inclined orbits the BP method has been used for estimation of the field on the segment of straight line perpendicular to the direction GO [78, 79]. The GIO transform (3.3.10), (3.3.26)-(3.3.28) is valid for general case of the inclined orbits. The GIO transform is not limited by the canonical transform requirements and the function $B(\eta)$ may be chosen to simplify Eqs. (3.3.26), (3.3.27), for example, $B(\eta) = 1$. One can find from relationships (3.3.10), (3.3.26) and (3.3.27)

$$E(p,\gamma) = I_c(y)E(y,z) = \int du A(u) B(\eta) \exp\{i\kappa[p\alpha - \Phi_j(u)]\}, \tag{3.3.29}$$

where $\eta = \sin(\alpha - \gamma)$, $\alpha = \sin^{-1} u$.

For this case the stationary phase (SP) method applied to (3.3.26), (3.3.27) gives the next connection between the angle α and parameter p

$$p = \frac{d\Phi_j(\sin\alpha)}{d\alpha} \quad . \tag{3.3.30}$$

In a homogeneous part of medium the function $\Phi_j(u)$ may be determined by use of (3.3.30)

$$\Phi_j(u) = y_j \sin\alpha + z_j \cos\alpha \ . \tag{3.3.31}$$

For this case the stationary phase method applied to (3.3.26), (3.3.27) gives a connection between α and p

$$p = -z_j \sin\alpha + y_j \cos\alpha \ . \tag{3.3.32}$$

Equation (3.3.32) defines p as the distance between the *j-th* ray and the centre of the co-ordinate system - point *O* (Fig. 3.3.1). If the centre of global spherical symmetry of the medium coincides with point *O*, p is the impact parameter of the *j-th* ray. For the evaluation of the GIO transforms in two considered partial cases it is not necessary to find the angular spectrum $A(u)$. The SP method gives the next formula for the transformed field

$$E(p,\gamma) = E_a(p,\alpha)\exp[ik\Phi_a(p,\alpha)],\ \Phi_a(p,\alpha) = p\alpha - z_j \cos\alpha - y_j \sin\alpha \ . \tag{3.3.33}$$

where $E_a(p,\alpha)$ is the coefficient describing contribution of the stationary point corresponding to the *j-th* physical ray. The ray direction angle α can be determined from (3.3.33) by differentiating the phase of the field $E(p,\gamma)$:

$$d\Phi_a(p,\alpha)/dp = \alpha(p) \ . \tag{3.3.34}$$

Then the bending angle $\xi(p)$ can be found from $\alpha(p)$ (3.3.34) by using equation (3.3.21) if the distance r_g is known. This is only possible if the orbit of the satellite G is circular, i.e. $r_g = const$. In the case of elliptical orbit one may apply the RHFSA method or the phase matching method considered in section 3.2 to find dependence of $\xi(p)$. When the modified refraction index $M(r) = \dfrac{n(r)r}{a}$, where a is the earth's radius, is a monotonic function, only one physical ray can correspond to the impact parameter p. A possibility of the multi-path effect corresponding to monotonic $M(r)$ profiles has been shown earlier in [69]. In this case the GIO can disentangle the multi-path rays expressing the ray direction angle $\alpha(p)$ as a single-valued function of the impact parameter p.

3.4. Combined amplitude-phase method

A relationship between the eikonal acceleration, Doppler shift, phase, and intensity variations of RO signals has been revealed from theoretical considerations and experimental analysis of RO radio-holograms [55, 82, 83]. The introduced eikonal acceleration technique seems to be simpler as compared with the thin phase screen model and radio-holographic back-propagation method [84, 85] proposed for location of plasma structures in the ionosphere. This technique may be applied also to estimate the total atmospheric absorption. The absorption measurements are planned for future RO missions [87] to determine water vapor abundance in the stratosphere and troposphere. A differential Canonical Transform/Full Spectrum Inversion (CT/FSI) technique was proposed to retrieve absorption in X/K band, 9–22 GHz [88]. The total absorption effect in the trans-atmospheric telecommunication link orbital station MIR – geostationary satellites was measured at frequency 930 MHz [30, 31]. In these experiments the refractive attenuation has been excluded by use of the phase and Doppler frequency data. A possibility to measure absorption in the atmosphere in X/K band by use of a spectral phase matching method (SPMM) and Fourier Integral Operators (FIO) has been detected in [76, 77]. In these works the second derivative of the phase on time was used for excluding the refractive attenuation from an amplitude function of the SPMM and FIO spectra. The eikonal acceleration technique can be directly applied to estimation of the total absorption in the atmosphere from analysis of RO data. There is a significant relationship between the phase path excess $\Phi(t)$ and amplitude $A(t)$ of radio-holograms considered as functions of time at two GPS frequencies f_1 and f_2. This relationship is described for the case of local spherical symmetry by Eqs. found formerly in [22, 53, 55, 65, 89–91]:

$$\Phi(p) = L(p) + \kappa(p) - R_0, \tag{3.4.1}$$

$$X(p) = pR_0[p_s d_1 d_2 |\partial\theta / \partial p|]^{-1}, \tag{3.4.2}$$

$$\partial\theta / \partial p = d\xi / dp - (1/d_1 + 1/d_2), \tag{3.4.3}$$

$$L(p) = (R_1^2 - p^2)^{1/2} + (R_2^2 - p^2)^{1/2} + p\xi(p), R_0 = (R_1^2 - p_s^2)^{1/2} + (R_2^2 - p_s^2)^{1/2} \tag{3.4.4}$$

$$d_1 = (R_1^2 - p^2)^{1/2},\ d_2 = (R_2^2 - p^2)^{1/2} \tag{3.4.5}$$

where $\kappa(p)$ is the main refractivity part of the phase path excess, $\xi(p) = -d\kappa(p)/dp$ is the refractive angle, $\theta(p)$ is the central angle, p, p_s are the impact parameter of the ray trajectory GTL, and the line of sight *GDL*, respectively, R_1, R_2, R_0 are the distances GL, OG, and OL, correspondingly, $L(p)$ is the distance GABL, and d_1, d_2 are two short lengths: GA and BL (Fig. 3.4.1). The refractive angle $\xi(p)$ is connected with the central angle $\theta(p)$ (Fig. 3.4.1):

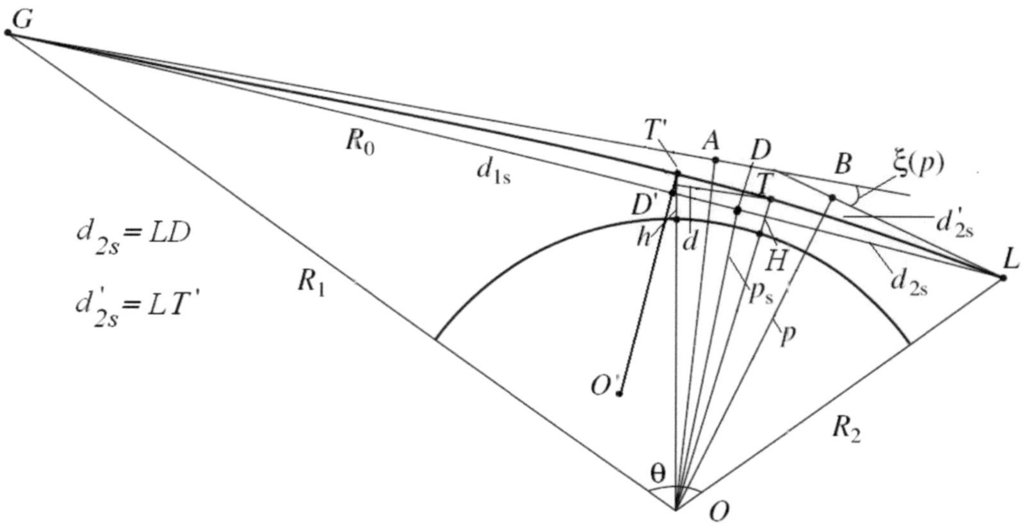

Fig.3.4.1. Ray path configuration in a spherically symmetric media during the RO experiment. Vertical gradient of the refractivity is perpendicular to the radio ray *GTL* in the tangent point *T*. In the case of the spherical symmetric atmosphere the tangent point *T* coincides with the radio ray perigee. The center of spherical symmetry *O* and the tangent point *T* can be displaced owing to the horizontal gradients in the atmosphere to point *O'* and *T'*, respectively.

$$\theta = \pi + \xi(p) - \sin^{-1}(p/R_1) - \sin^{-1}(p/R_2),\ \theta = \pi - \sin^{-1}(p_s/R_1) - \sin^{-1}(p_s/R_2), \quad (3.4.6)$$

$$\xi(p) = \sin^{-1}(p/R_1) - \sin^{-1}(p/R_2) - \sin^{-1}(p_s/R_1) - \sin^{-1}(p_s/R_2). \quad (3.4.7)$$

Two different expressions (3.4.6) for the central angle $\theta(p)$ can be used to obtain relation between the impact parameters p_s and p, and the refractive angle $\xi(p)$. From Eqs. (3.4.1) and (3.4.6), one can obtain by differentiation of the phase excess $\Phi(p)$ and the central angle $\theta(p)$ over time the relationship connecting the Doppler frequency of RO signal with the impact parameters p_s and p:

$$d\Phi(p)/dt = (p - p_s)d\theta/dt + (d_1 - d_{1s})R_1^{-1}dR_1/dt + (d_2 - d_{2s})R_2^{-1}dR_2/dt, \quad (3.4.8)$$

$$d\theta/dt = \partial\theta/\partial p_s\ dp_s/dt - p_s[(d_{1s}R_1)^{-1}dR_1/dt + (d_{2s}R_2)^{-1}dR_2/dt]$$
$$d_{1s} = (R_1^2 - p_s^2)^{1/2}, d_{2s} = (R_2^2 - p_s^2)^{1/2} \quad (3.4.9)$$

where d_{1s}, d_{2s} are the distances GD and DL, correspondingly, (Fig. 3.4.1). After substitution of the equation (3.4.9) in equation (3.4.8) one can obtain:

$$d\Phi(p)/dt = (p-p_s)\{(1/d_{1s}+1/d_{2s})dp_s/dt-(p^2-p_s{}^2)$$
$$\{R_1 dR_1/dt[d_{1s}(d_{1s}+d_1)(pd_{1s}+p_s d_1)]^{-1}$$
$$+R_2 dR_2/dt[d_{2s}(d_{2s}+d_2)(pd_{2s}+p_s d_2)]^{-1}\}\} \quad (3.4.10)$$

Eqs. (3.4.6) – (3.4.10) are valid for the general case of the non-circular orbits of the GPS and LEO satellites. Equation (3.4.10) can be simplified under condition:

$$\left|(p^2-p_s{}^2)\{R_1 dR_1/dt[d_{1s}(d_{1s}+d_1)(pd_{1s}+p_s d_1)]^{-1}+R_2 dR_2/dt[d_{2s}(d_{2s}+d_2)(pd_{2s}+p_s d_2)]^{-1}\}\right|$$
$$<< \left|1/d_{1s}+1/d_{2s})dp_s/dt_s\right| \quad (3.4.11)$$

which is valid when $\left|(p-p_s)dR_{1,2}/dt\right| << \left|p_s dp_s/dt_s\right|$. This inequality is valid for all RO situations. Under the indicated approximation one can obtain from equation (3.4.10) a formula for estimation of the difference $p-p_s$ from the Doppler shift $F_d = d\Phi(p)/dt$:

$$d\Phi(p)/dt = F_d = -(p-p_s)(1/d_{1s}+1/d_{2s})dp_s/dt \quad (3.4.12)$$

The values p_s, dp_s/dt, d_{1s}, d_{2s} can be derived from orbital data, and the phase delays $\Phi_{1,2}(p)$ are the objects of measurements and are given in the phase parts of radio-holograms at frequencies f_1 and f_2. Therefore, one may obtain (after the ionospheric correction) dependence of the refractive angle $\xi(p)$ on the impact parameter p and then use the Abel inversion to retrieve vertical profiles of the refractivity $N(h)$, pressure $P(h)$, temperature $T(h)$, humidity $e(h)$, and electron density in the ionosphere $N_e(h)$ [16, 52].
A new relationship connects the eikonal acceleration and refractive attenuation of radio waves. Under conditions

$$|p-p_s| << p_s, \quad (3.4.13)$$
$$\left|(p-p_s)d(\partial\theta/\partial p_s)/dt\right| << \left|(dp/dt-dp_s/dt)\partial\theta/\partial p_s\right| \quad (3.4.14)$$

by use of equation $dp/dt-dp_s/dt \approx [X(t)-1]dp_s/dt$ [22,55,83], one can obtain from (3.4.12):

$$1-X(t) = ma_\Phi = m\frac{d(\Delta f)}{dt} = m\frac{d^2\Phi}{dt^2}, \quad (3.4.15)$$
$$m = q/(dp_s/dt)^2, \quad (3.4.16)$$
$$q = d_{1s}d_{2s}/(d_{1s}+d_{2s}),\ d_{1s}+d_{2s} = R_0. \quad (3.4.17)$$

Equation (3.4.15) indicates equivalence between variations of the phase path excess (eikonal) acceleration $a_\Phi = \frac{d^2\Phi}{dt^2}$ and refractive attenuation $X(t)$, which may be verified by the RO data. Usually during RO experiments parameters m and dp_s / dt are known from the orbital data because of location of the spherical symmetry centre O and its projection on the line of sight - point D are known (Fig. 3.4.1). Therefore, equation (3.4.15) gives a possibility to recalculate the eikonal acceleration a_Φ to the refractive attenuation X_a. This is useful for excluding systematic errors from the phase and/or amplitude data. This is also useful for estimating the total absorption in the atmosphere. The refractive attenuation X_a is determined from the amplitude data as a ratio of the intensity of radio signal propagating through the atmosphere $I_a(t)$ to its intensity in free space I_s :

$$X_a(t) = I_a(t) / I_s . \tag{3.4.18}$$

Thus, the experimental value X_a is the sum of the refractive and absorption contributions. However, the eikonal acceleration depends on the refraction effect only. This gives a possibility to determine the absorption in the atmosphere $Y(t)$ as a ratio:

$$Y(t) = \frac{X_a(t)}{1 - ma_\Phi} . \tag{3.4.19}$$

This possibility should be investigated in detail because in the future satellite RO missions measurement of the absorption effects due to water vapor and minor atmospheric gas constituents is planned and the difficulties consist in removing the refractive attenuation effect from the amplitude data. Equation (3.4.19) indicates a feasible way to solve this problem. Also Eq. (3.4.19) may be useful for estimating conditions for communication in the Ku/K bands between two LEO satellites in a radio occultation geometry [86]. At the GPS frequencies, the total absorption effect in RO experiments caused mainly by the atmospheric oxygen is about 1-3 dB in the lower troposphere and is small in the upper troposphere and stratosphere.

The horizontal gradients in the ionosphere and atmosphere can displace the tangent point T and centre of spherical symmetry O from its standard positions to points T′ and O′, respectively (Fig. 3.4.1). As a consequence, value of distance T′L (approximately equal to LD′ (Fig. 3.4.1)) will change to d'_{2s}, and parameter m will also change its magnitude. However, Eqs. (3.4.15)-(3.4.17) are valid in the case of local spherical symmetry with new centre O′. Therefore, if the magnitude of parameter m could be estimated from experimental data, then it is possible to find new value of the distance T′L= d'_{2s}, and to determine location of new tangent point T′ relative to point T (or L). Parameter dp_s / dt can be found from trajectory data, which describes motions of the GPS and LEO satellites relative to the center of spherical symmetry - point O (Figure 3.4.1):

$$dp_s / dt = v + (w - v) d_{1s} / R_0 , \tag{3.4.20}$$

where w, v are the velocity components of the GPS and LEO satellites, respectively. These velocity components are perpendicular to the straight line GL in the plane GOL. The components v and w are positive when oriented in the direction towards the point O and are negative in the opposite case. Eqs. (3.4.15)-(3.4.17), (3.4.20) give a way to find the distance LT d_{2s} from simultaneous observation of the phase and intensity variations:

$$d_{2s}' = 2mv^2[1 + 2mvR_0^{-1}(v-w) + (1 - 4mvwR_0^{-1})^{1/2}]^{-1}. \quad (3.4.21)$$

Therefore, if the magnitude of parameter m is estimated from experimental data, it is possible to find new value of distance T′L= d_{2s}' and thus determine location of new tangent point T′ relative to the point T (or L).
For determination of the parameter m from experimental data, we can assume that value m is a slowly changing function of time. If noise is very small, averaging is not necessary and parameter m can be determined directly from equation (3.4.15) as a ratio:

$$m = \frac{1 - X(t)}{a_\Phi}. \quad (3.4.22)$$

In the presence of noise, value $m(t_k)$ corresponding to some instant of time t_k can be determined from RO data as a ratio of the average of squared refractive attenuation and eikonal acceleration variations:

$$m(t_k) = \frac{\sum_{i=k-M}^{i=k+M} [1 - X(t_i)]^2}{\sum_{i=k-M}^{i=k+M} [a_\Phi(t_i)]^2}, \quad (3.4.23)$$

where $2M+1$ is the number of samples for averaging, $X(t_i)$, $a_\Phi(t_i)$ are the current values of the refractive attenuation and eikonal acceleration variations at the time instant t_i. Parameter m can be determined also from a correlation relationship:

$$m(t_k) = \frac{\sum_{i=k-M}^{i=k+M} [1 - X(t_i)] a_\Phi(t_i)}{\sum_{i=k-M}^{i=k+M} [a_\Phi(t_i)]^2}. \quad (3.4.24)$$

Relationship (3.4.24) describes parameter m as a correlation coefficient between the refractive attenuation $X(t_i)$ and eikonal acceleration $a_\Phi(t_i)$.
When parameter m is known with sufficient accuracy one can estimate the distance d between the tangent points T and T′ (Fig. 3.4.1) by use of Eqs. (3.4.16) and (3.4.17).

$$d \approx d_{2s}' - d_{2s} = d_{2s}' - (R_2^2 - p_s^2)^{1/2}. \quad (3.4.25)$$

With influence from turbulence, multi-path and instabilities of the transmitter and receiver become important Eqs. (3.4.22) – (3.4.25) give three different values of the parameter m and distance d . Note that one should account for the velocity of the point O′ in relation to (3.4.16) and (3.4.17). However in the case of the earth's atmosphere and ionosphere, the velocity of motion of layers, and consequently, the velocity of point O′ are small as compared to the orbital velocities of the GPS and LEO satellites and its contribution may be neglected.

Below, examples of application of suggested method are given by analysis of the FORMOSAT3 RO events. An example of determination of the displacement d is demonstrated in Fig. 3.4.2. The data shown in Fig. 3.4.2 correspond to the FORMOSAT-3 RO event 0106, April 23, 2006, 04 h 11 m LT, with geographical coordinates 11.1 N 310.7 W. Curves 1 and 2 in Fig. 3.4.2, left panel, demonstrate significant correspondence between the refractive attenuations X_p and X_a , which are evaluated from the eikonal acceleration and amplitude data (Eqs. (3.4.23) and (3.4.24), respectively) at the first GPS frequency f_1. This correspondence allows one to determine the horizontal displacement d of the tangent point T. The results of the evaluation of the displacement d by use of Eqs. (3.4.21), (3.4.23) and (3.4.24) are shown in Fig. 3.4.2, right panel. Curves 1 and 2 correspond to values d found by the relationships (3.4.21), (3.4.23); and (3.4.21), (3.4.24), respectively. According to Fig. 3.4.2 (right panel) the displacement d estimated from Eqs. (3.4.23) and (3.4.24) is bounded between ± 25 km

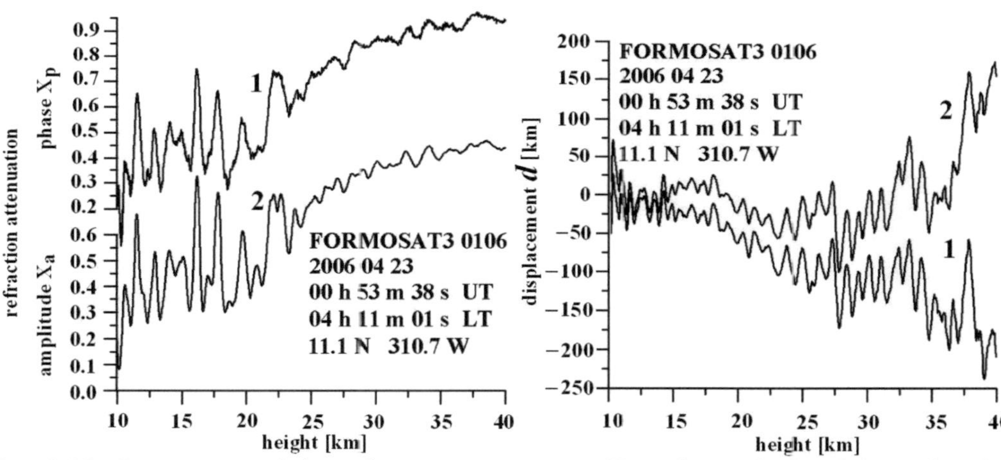

Fig. 3.4.2. Comparison of the refractive attenuations X_a and X_p calculated from the phase (curve 1) and amplitude (curve 2) data (left). Displacement d of the tangent point calculated by use of Eqs. (15) and (16) (curve 1 and 2, respectively) (right).

in the 10 km - 16 km altitude interval and between ± 50 km in the 16 km - 35 km height interval. This experimental result open possibility to apply suggested method for location of layered structures in the atmosphere.

3.5. Concluding remarks

This section has considered the radio-holographic methods in different areas of application. The RHFSA method is optimal for compression of the angular spectrum of RO signals with aim to visualize weak reflections from the earth's surface, to detect multi-path propagation, and obtain 1-D images of the atmosphere and/or ionosphere. To achieve this aim the RHFSA method may be applied under minimal limitations as compared to other methods. The RHFSA method may be applied for analysis of RO data in the cases of the multi-path propagation, horizontal gradients, multi-valued connection between the bending angle and the impact parameter, existence of turbulence. However the technical difficulties exist for application of the RHFSA method when finding dependence of the bending angle on the impact parameter because an atmospheric model in the region of the RO sounding is required for constructing the reference signal . Model of the reference signal may be elaborated, in principle, for different cases of the single-ray and multi-path propagation, including waveguide propagation, and in presence of the horizontal gradients and caustics. Therefore the method is the most general tool for revealing layered structures and determination of vertical profiles of physical parameters in the atmosphere and ionosphere. A necessity of solution of direct problem for elaborating of the reference signals by use of given atmospheric model is a shortcoming of the RHFSA method. In the lower layers of the atmosphere, it is necessary to find appropriate values of model parameters for the most effective compression of the spatial spectrum. This introduces difficulties in the RO data analysis especially in conditions of multi-path propagation. Advantages of the RHFSA method were demonstrated during measurements of the characteristics of the signals reflected from the earth surface [71–74], and, also, during analysis of the multi-path propagation in the lower layers of the atmosphere [65, 71, 74].

The phase matching method [75–77] successfully uses the analytical form of the geometrical part of the phase path for elaborating the reference signal. This analytical form is common to all atmospheric spherical-symmetric layered structures and it allows one to obtain directly by the differentiation of the phase of the transformed radio-hologram dependence of the bending angle on the impact parameter. The phase matching method and RHFSA method are applicable in general case of elliptical orbits. The integration is provided in these methods along the actual orbital trajectory of the satellite. This excludes the need to recalculate the field to the straight lines which would otherwise be necessary when applicate GIO and CT methods based on the Fourier Integral Operators (FIO). Phase matching method is functioning under condition of the spherical symmetry of the atmosphere, and single-valued connection between the impact parameter and the bending angle. Important advantage of the phase matching method consists in automatic excluding of the refractive attenuation of radio waves from spectrum of the RO radio-hologram. This opens up a possibility of precise measurements of the total absorption of radio waves during RO experiments in the water vapor and other minor atmospheric gas constituent absorption lines at the millimeter and centimeter wavelength bands [87, 88]. The phase matching method currently is a basic technique for operative RO data analysis. The CT method [78, 79, 81] appeared before the phase matching method and has been broadly used for determining dependence of the bending angle on the impact parameter. The CT method requires recalculating the field to a straight line from the orbital trajectory. This recalculation is conducted by the use of the BP method. The spectrum of the CT transform may also be applied to measurements of the total absorption of radio waves . Main condition for the application of the CT method is the spherical symmetry of the atmosphere and single-valued connection between the impact parameter and bending angle. When the modified refraction index $M(r)$ is a monotonic function, only one physical ray may correspond to the impact parameter p. A possibility of the multi-path effect corresponding to monotonic $M(r)$ profiles has been shown earlier in [69]. In this case the GIO, CT, and phase matching method can disentangle the multi-path rays expressing the bending angle as a single-valued function of the impact parameter p.

The amplitude-phase method [55, 82, 83, 89–91] may be applied in the case of one ray propagation. This is important for identification of layered structures in the atmosphere and ionosphere, and estimation of total absorption in the atmosphere. Based on forty years of radio occultation (RO) experiments it is now recognized that the eikonal acceleration of radio waves (equal to the time derivative of the Doppler shift), derived from analysis of high-stability Global Positioning System (GPS) RO signals, is as important as the Doppler frequency. The eikonal acceleration technique allows one to convert the phase and Doppler frequency changes into refractive attenuation variations. From such derived refractive attenuation and amplitude data, one can estimate the total absorption of radio waves. This is important for future RO missions when measuring water vapor and minor atmospheric gas constituents because the difficulty of removing the refractive attenuation effect from the amplitude data can be avoided. The eikonal acceleration technique can be applied also for determining the location and inclination of sharp layered plasma structures (including sporadic E_s layers) in the ionosphere. Advantages of the eikonal acceleration technique are validated by analyzing RO data from the Challenging Minisatellite Payload (CHAMP) and the FORMOSA Satellite Constellation Observing Systems for Meteorology, Ionosphere, and Climate missions (FORMOSAT-3/COSMIC).

4. Results of RO investigation of the earth's atmosphere

4.1. Atmospheric variations of RO signals and accuracy of determination of vertical profiles of the refractivity and temperature

In the preceding chapters a possibility of restoration of vertical profiles of pressure and temperature in the atmosphere from variations of parameters of radio signals has been substantiated. As it was noted earlier, for determination of these characteristics it is enough to find connection of the bending angle and impact parameter from measurements of the frequency, phase or amplitude. In this chapter examples of RO signals variations caused by the atmospheric influence are adduced and step-by-step transfer from measurements to determination of the atmospheric parameters is considered . Measurements errors and inaccuracies of the used algorithms of data inversion influence on the retrieved values of atmospheric characteristics, therefore, in this chapter a short description of the basic errors sources is conducted also. Values of the atmospheric parameters, determined by RO method, are compared with results, obtained by other technical means. RO sounding of the atmosphere allows obtaining information not only about the above mentioned characteristics, but also about the wave, layered and turbulent structures in the atmosphere, a possibility of their research by the RO method also will be considered in this chapter.

Before description of RO technique of the atmospheric data revealing, let us adduce results of the first RO experiments, conducted in the decimeter ($\lambda_1 = 32$ cm) and centimeter ($\lambda_2 = 2$

cm) ranges. During these experiments the monochromatic radio waves were emitted by the MIR station transmitters and were received onboard geostationary satellites that retranslated the received signals without any change of their structure to the ground based measurement station. The distances from transmitter and receiver up to the radio ray perigee (point T in Fig. 2.2.2) were equal to 42000 km and 2,100 km, respectively, vertical velocity of intersection of the atmospheric layers by the straight line satellite-satellite was contained in different experiments in the interval 1.4 – 2.1 km s^{-1}. Transmitters and receivers of the satellites do not be developed specially for RO sounding, however because the broad diagrams of transmitting and receiving antenna and expanded dynamical range of the receivers the RO signal reception was provided up to contact of the radio ray with the earth's surface. Note that the relative accuracy of restoration of vertical profiles of the atmospheric parameters from the amplitude measurements is far below corresponding values relevant to the phase data. Despite this variations of the signal amplitude give good visualization of vertical atmospheric structures including the layered, turbulent or wave formations. This difference appeared because the amplitude variations are inversely proportional to the derivative of the bending angle with respect to the impact parameter; however the phase path excess is proportional to the refractivity. Besides this the amplitude variations depend on vertical size of the Fresnel' zone, however the eikonal in the neutral atmosphere does not depend on the wave length.

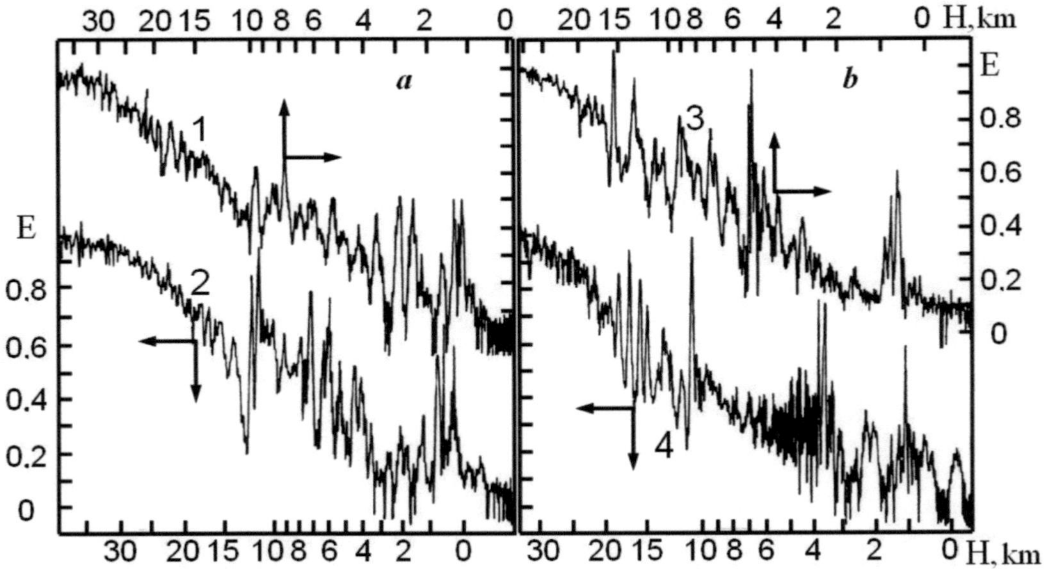

Fig. 4.1.1.Typical variations of the field strength E of the radio waves in the range $\lambda_1 = 32$ cm. Curves 1, 3, 4 are relevant to radio occultation above the Indian Ocean in December 1990. Curve 2 corresponds to radio occultation above the Kazakhstan in June 1990.

In Fig. 4.1.1 typical variations of the field strength E of radio waves in the range $\lambda_1 = 32$ cm obtained from RO data [28, 29] are shown as functions of the minimal height of radio ray H. The field strength E is normalized relative to the level corresponding to free space. Dependences $E(H)$, shown by curves 1, 3, 4; and curve 2 were obtained during radio occultation above the Indian Ocean in December 1990, and above the Kazachstan in June 1990, respectively. Analogous dependences $E(H)$ were obtained at the wave length $\lambda_2 = 2$ cm. Determination of the altitude H has been provided by use of satellites trajectory data and a model of the altitude dependence of the atmospheric refractivity. Usage of model may lead to an error in the altitude measurements in the lower troposphere nearly equal to 1 km. Since the decimeter radio waves practically are not absorbed in the atmosphere, then variations of the field strength are caused mainly by the refractive attenuation . Measurements showed that a notable regular decreasing of the signal amplitude is beginning at the altitude nearly equal to 30 km. Despite difference of conditions in the investigated regions located above the land or ocean and distinct seasons of year, dependences $E(H)$ have many common features. The tropopause influence is notable on all RO curves and become apparent as a sharp decreasing and next increasing of the signal amplitude, all vertical profiles $E(H)$ demonstrated a complex vertical structure of the refractivity in the troposphere. At the altitude of the radio ray perigee H=5 km the amplitude of radiowaves is changing in the interval 0.3–0.48 depending on RO region. At the altitude H= 2 km the refractive attenuation may change from 0.16 up to 0.27 with the average value equal to 0.21 . Measured near the earth's surface values of the field strength E are about two times below the theoretical magnitudes. This distinction may be caused by a diffraction phenomenon. Radius of the first Fresnel' zone was equal nearly to 0.5 km in RO experiments and therefore at the altitude H<1 km the surface influence may be significant. Radio field near the surface may be attenuated owing to the refraction and diffraction effects. Complex structure of altitude profile of refractivity in the troposphere, caused by the temperature inversions, spatial and temporal variations of the humidity distribution, and also by meso-scale irregularities, cloudiness, and atmospheric fronts, leads to significant variations of the radio waves amplitude. Analysis of the refractive attenuation was conducted in [28, 29] by use of real profiles of the meteorological- parameters and models of the atmospheric structure and it was shown, that the amplitude variations of RO signal passed through the atmosphere follow altitude location and intensity of the refractivity features. Layered structures, for example, the clouds, in which the temperature is depressed relative to the background, and therefore the refractivity is by several N-units (1 N-units=10^{-6}) greater the background values, are leading during immersion of radio ray in the atmosphere at first to decreasing, and then to increasing of RO signal amplitude. A layer having the increased temperature causes the increase of the amplitude when the minimal height of radio ray H is decreasing with the next decreasing to a level below the average one in a given altitude interval. Intensity of variations of the refractive attenuation may be characterized by the magnitude of the difference between the local maxima and minima of the amplitude, and their vertical extension is proportional to the depth of the temperature inversion and the width of the inversion layer.

Analogously dependences of the refractive attenuation were obtained at wave length λ=19 cm by use of the satellites MICROLAB and CHAMP. Examples of variations of the refractive attenuation as functions of the height H are given in Fig. 4.1.2. Along the horizontal axis values of the minimal radio ray height H are marked. Data in Fig. 4.1.2 correspond to a tropical region. The altitude interval 5 km– 30 km is shown in the left panel. The mesosphere and lower ionosphere altitudes are demonstrated in the right panel in Fig. 4.2.1. Curve 1 and curve 2 correspond to the refractive attenuations found from the amplitude and phase data. Curves 3 are relevant to the theoretical model of the refractive attenuation. The sampling rate during RO measurements was equal to 50 Hz; RO measurements began at the minimal height of radio ray nearly to 130 km and ending after disappearance of signal. Duration of typical RO session, conducted by the CHAMP satellite, having the orbit with the height 470 km, was in average equal to nearly 100 s. The amplitude variations caused by the ionospheric influence are seen in the beginning of measurements shown in Fig. 4.1.2 (right). After the ionospheric variations when the height of radio ray is decreasing at first the insignificant amplitude variations of signal are observed near the average level, transferring into the regular decreasing at the altitudes below 30 km. In the end of the RO session the intensive amplitude variations are seen because of the large vertical gradients of refractivity near the tropopause at the altitude $H \approx 17$ km typical of tropical regions.

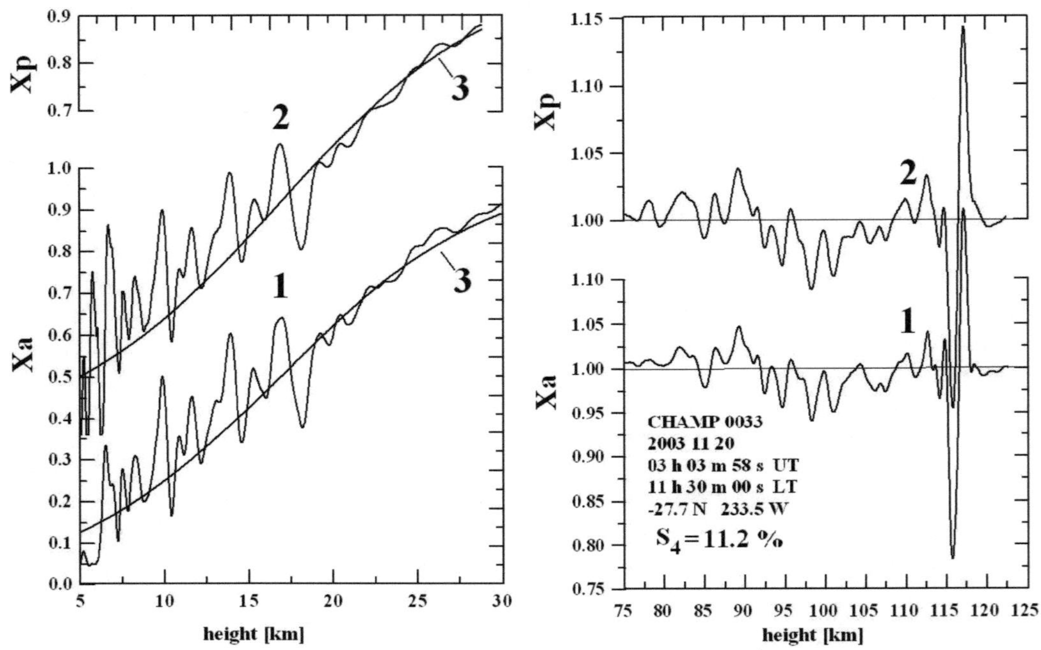

Fig. 4.1.2. Examples of vertical profiles of refractive attenuation at λ =19 cm.

Method of obtaining of information about the amplitude and Doppler shift variations caused by the atmospheric influence depends on the structure of RO signals and the technical possibility of measurements of their parameters. The large number of the operating navigational satellites, emitted high-stable signals indicated a good possibility of their further application for the global monitoring of the atmosphere. Usage of the navigational satellites GPS, GLONASS or GALILEO as emitters of radio waves strictly determines the frequency range of emitted signals, their spectrum, power, amplitude and phase stability. In the present time during RO experiments measurements of the amplitude and phase of RO signals are conducted at frequencies f_1 = 1575.42 MHz and f_2 =1227.6 MHz. The amplitude of RO signal is measured as a voltage ratio of signal to noise, a phase of signal is determined as the phase path excess (eikonal) of radio waves. Therefore the amplitude of RO signal is proportional to the strength of electromagnetic field, it depends

on receiver parameters: the noise temperature, the gain coefficient of the receiving antenna, and ability of a spherical lens – the earth's atmosphere - to focus (or defocus) radio waves.
As follows from the scheme of restoration of the atmospheric parameters indicated in the Table 2.3.1 after measurements of the amplitude and phase path excess as functions of time t, it is necessary to determine dependences of the bending angle $\xi(p)$ on the impact parameter p. For this aim one may use separately the amplitude and phase data. The most accurately measured physical parameters are the time, frequency and phase of oscillating or wave process therefore usually the phase data are used for atmospheric parameters restoration. To find the bending angle it is necessary to determine the Doppler shift of RO signal Δf, caused by atmospheric influence.The Doppler shift may be determined from the relationship

$$\Delta f = \lambda^{-1} d\Phi / dt\,. \tag{4.1.1}$$

From known Doppler shift the bending angle and impact parameter may be determined as functions of time $\xi(t)$, $p(t)$ by use of trajectory data. The found functions $\xi(p)$ and p are caused by influence of the neutral atmosphere and ionosphere. Quick change of the phase path excess in the beginning of RO session is caused by propagation of radio waves through the ionosphere. Influence of the ionosphere is an interfering factor for determining the atmospheric characteristics.
The temperature restoration at the altitudes below 40 km with accuracy nearly 1 K requires knowledge of the neutral atmosphere contribution in the bending angle up to the 100 km altitude. The ionosphere introduces at the altitudes above 50 km a basic contribution in the eikonal and, respectively, in the bending angle. The neutral atmosphere contribution prevails below the altitude 40 km [67]. For determination of the neutral atmosphere contribution in the bending angle it is necessary to exclude the ionospheric component. Removing of the ionospheric component usually is accomplished by different ionospheric correction methods. The simplest method of ionospheric correction consists in exclusion of the ionospheric component in the eikonals Φ_1 and Φ_2 by use of relationship

$$\Phi = \left(f_1^2\Phi_1 - f_2^2\Phi_2\right)\left(f_1^2 - f_2^2\right)^{-1}, \tag{4.1.2}$$

where Φ is the phase path excess in neutral medium, f_1 and f_2 are the frequencies of the radio waves emitted by GPS system, Φ_1 and Φ_2 are measured values of the phase path excess. At the altitudes below 40 km values Φ_1 and Φ_2 are practically the same and this allows neglecting interfering influence of the ionosphere at these heights. At large heights application of the Eq. (4.1.2) does not always ensures sufficient correction of the ionospheric influence and this may leads to systematic errors in determination of the atmospheric characteristics. This is connected with different paths of propagation of radio waves of the ranges f_1 and f_2 because of dispersive properties of the ionospheric plasma. These paths are characterized by the different impact parameters. Analysis conducted in [37] showed, that difference of the altitudes of the ray perigee in the ranges f_1 and f_2 may change in the 10–500 m interval depending on the current state of the ionosphere and the height of sounding in investigated region. For decreasing of the errors in determination of the atmospheric parameters, connected with the difference of the propagation paths of radio waves having different frequencies, another ionospheric correction method was suggested in [92]. This method consists in separate determining dependences of the bending angles ξ_1 and ξ_2 on the impact parameter p for radio waves having the frequencies f_1 and f_2 and presentation of dependences of atmospheric contributions to the bending angle $\xi(p)$ in the form of a linear combination

$$\xi(p)=\left[f_1^2\xi_1(p)-f_2^2\xi_2(p)\right](f_1^2-f_2^2)^{-1}. \quad (4.1.3)$$

Since values ξ_1 and ξ_2 are determined with the constant time intervals, the interpolation of dependences ξ_1 and ξ_2 is necessary for ionospheric correction by use of Eq. (4.1.3). Note that any method of the ionospheric correction does not exclude at all the interfering influence of the ionosphere and, in particular, the fluctuations of the bending angle, caused by the ionospheric irregularities. Comparison of found after the ionospheric correction bending angle ξ with values, calculated from meteorological model of the atmosphere, showed, that caused by the ionospheric irregularities relative error in determination of ξ is large even at the minimal height of radio ray equal to 50 km [67]. At the altitudes 70–100 km unavoidable variations of ξ are similar to homogeneous high-frequency noise [67]. Vertical profiles of the bending angle ξ in the upper atmosphere are near to the exponential one and in the altitude range 70–100 km values ξ should diminish from 10^{-6} up to 10^{-8} radian. In the same time according to results of RO measurements values of ξ are changing in the interval from 10^{-5} to 10^{-6} radians. The remaining error of ionospheric correction arises owing to influence of the small-scale ionospheric irregularities. Due to the different propagation paths of RO signals at frequencies f_1 and f_2 the effect of influence of the small-scale irregularities becomes uncorrelated and application of the relationship (4.1.3) only reinforces the uncorrelated noises, existing in two channels of the phase measurements. Diminishing of the interfering ionospheric influence may be achieved only by sounding of the atmosphere at the smaller wavelengths, for example, in the centimeter range. After consideration of possibilities of the existing technical tools another methods were suggested for decreasing of the ionospheric influence on the results of restoration of the atmospheric parameters. One of them is the statistical regularization method consisted in changing of the experimental values of the bending angle ξ at the altitudes $H>$ 45 km by values ξ, calculated by use of a radio meteorological model of the atmosphere [67]. Analysis of the errors in measurement of bending angle and their influence on the accuracy of restoration of vertical profiles of the atmospheric parameters has been conducted in [93]. In this work it was shown, that, since at the 50 km altitude of the radio ray perigee $\xi \approx 10^{-5}$ radians, at the 35 km altitude $\xi \approx 10^{-4}$ radians, and near the earth' surface $\xi \approx 10^{-2}$ radians, the relative error in the ξ measurements is quickly diminishing with the height decrease.

Before description of the next stages of atmospheric parameters determination, consider variations of the bending angle ξ under different conditions of the atmospheric sounding.

The bending angle ξ determined from the phase measurements, in essence, is an independent integral parameter, characterizing the atmospheric state. The ionosphere practically does not influence on the accuracy in measurements of ξ at the altitudes below 35 km. Therefore results of measurements of the bending angle, conducted in the different frequency ranges, in different time and in different climatic zones and having small relative errors, may be useful for detecting of the long-term trends in the atmospheric state.
The next stage after finding the bending angle and the impact parameter– in correspondence with the table 2.3.1 - is determining the altitude profile of the refractivity $n(r)$ by use of the Abel inverse transformation (2.3.4) and vertical profiles of the atmospheric parameters in correspondence with Eqs. (2.3.8– 2.3.10). Before description of these stages of the data analysis it is necessary to study the sources of RO method errors. Before conducting the regular RO experiments in the work [37] an analysis of the different sources of measurement errors, inaccuracies of data handling and their influence on the accuracy of restoration of the atmospheric parameters was conducted. These sources may be unified in three groups: measurement errors, the calibration inaccuracies and errors of restoration of the atmospheric parameters. Measurement errors are caused by the technical inaccuracies: values of the ratio signal to noise, phase instability of the transmitter and receiver oscillators, clocks instability, multi-path effect and variations of the phase centers location of the transmitting and receiving antenna. The calibration errors depend on the interfering influence of the ionosphere and accuracy of determination of the satellites orbital parameters. Calibration errors are individual for each RO mission and in each measurement session. The common sources of errors influencing on the results of RO measurements are: errors in the constant values included in the expression for the refractivity (2.1.3); errors, connected with the upper boundary conditions in the Abel transformation (2.3.4) and in Eqs. (2.3.8) and (2.3.10); horizontal gradients of the refractivity and uncertainties in the altitude distributions of the water vapor in investigated region. In this section the sources of all mentioned errors do not be considered in detail because their extended description is already given in the work [37].
For determination of the altitude profile of the refraction index $n(r)$ and vertical profiles of the atmospheric parameters it is necessary to set the initial values of the bending angle ξ_0 or the impact parameter p_0 at the altitude h_0 below which the atmospheric characteristics are restored, and the corresponding to this altitude initial values of gas density, temperature T_1 and refractivity N_1. Inaccuracy in setting of these initial values are leading to the systematic deviations in restoration of the altitude profile of temperature $T(h)$; these errors are large near the upper boundary h_0 and are diminishing when the height is decreasing. The estimations conducted in [67] showed, that the 20% error in setting an initial value of the bending angle at the upper boundary at the altitude $H_0 \approx 50$ km leads to the temperature errors near 2 K at the height 20 km. The initial values do not influence on the results of the temperature restoration at the altitudes below 40 km if they are setting at the height about of 90 km.

4.2 Examples of evaluations of the atmospheric parameters from preliminary results of FORMOSAT-3/COSMIC RO mission

The geographical distribution of the first 10-day FORMOSAT-3/COSMIC RO mission events is given in Fig. 4.2.1. from April 21 to 30, 2006. The total number of the RO events is

812 because the main goal of the preliminary phase consists of checking the receivers software and hardware on board the FORMOSAT-3 satellites. After ending the preliminary phase (about 20 months after the launch) the nominal value of the RO events will achieve about 2500 occultations per day. Near polar orbit of the FORMOSAT-3 satellite allows global monitoring including North and South Polar regions, and ocean areas. As a rule, the first FORMOSAT-3/COSMIC RO measurements have been provided down to 5–8 km height interval. Tracking firmware for COSMIC receivers has been implemented by JPL [94]. L1 and L2 signals have been recorded in closed-loop mode above ~10 km. Below ~10 km L1 has been recorded in open-loop mode. While measuring the neutral atmospheric parameters, the ionospheric correction method is applied to subtract the ionospheric effect from the RO phase data. The linear ionospheric correction of the phase delays is applied automatically in the FORMOSAT-3/COSMIC RO data using formula (4.1.2) [95, 96]. The common bending angle correction method [92] requires differentiation of the phase excess on time and therefore enlarges the influence of high-frequency noise. Disadvantage of both linear correction methods consists of increasing of the high-frequency noise contribution. To diminish the high-frequency noise contribution, we exclude separately slow ionospheric trend from both frequencies f_1 and f_2 by use of the ionospheric model IRI-2000.

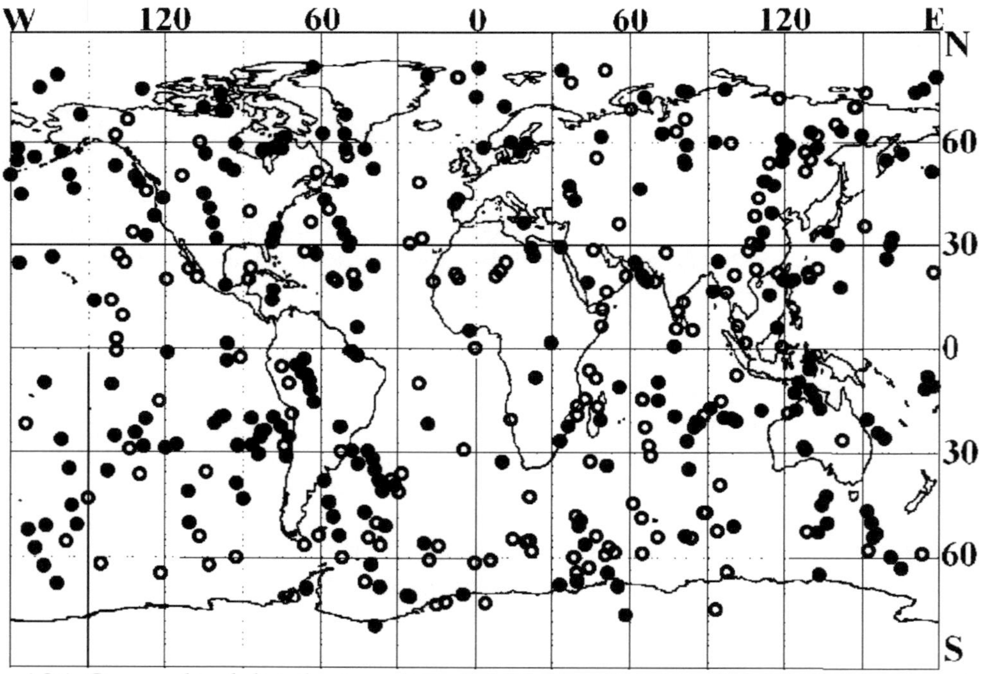

Fig. 4.2.1. Geographical distribution of the first 10-day FORMOSAT-3/COSMIC RO sessions from April 21 to 30, 2006. Open circles correspond to the night events from 20-00 to 08-00 LT. Closed circles correspond to the day events from 08-00 to 20-00 LT.

To eliminate a possible bias that model may introduce into the neutral refractivity profile we use the reference function which has been obtained from the phase delay $\Phi(t)$ by the least squares method. By this way of the ionospheric correction we exclude the ionospheric systematic error in the low frequency interval and increase in the high-frequency noise. The suggested method is appropriate during minimum in solar activity when the ionosphere is quiet. The results of application of the ionospheric correction to both phase delays f_1 and f_2 and comparison with the linear correction method (4.2.1) are shown in Fig. 4.2.2. The results, shown in Fig. 4.2.2, have been obtained after subtracting the phase delays, calculated using model of standard refraction in the atmosphere (2.1.4). The curves 1–3 in Fig. 4.2.2 (right) demonstrate the phase delays $\Phi_1(t)$ (curve 1), $\Phi_2(t)$ (curve 2) and linear corrected phase delay $\Phi(t)$ as function of the height h in the atmosphere. The noise from frequency $\Phi_2(t)$ is seen also in the phase delay $\Phi(t)$ (curve 3). Application of the ionospheric model allows excluding ionospheric effect from both phase delays $\Phi_1(t)$ (curve 1) and $\Phi_2(t)$ (curve 2) without increasing the noise intensity (Fig. 4.2.2, left). For better comparison with results of linear ionospheric correction (curve 3) curves 2 and 3 are displaced by 0.5 m in the opposite sites (Fig. 4.2.2, left).

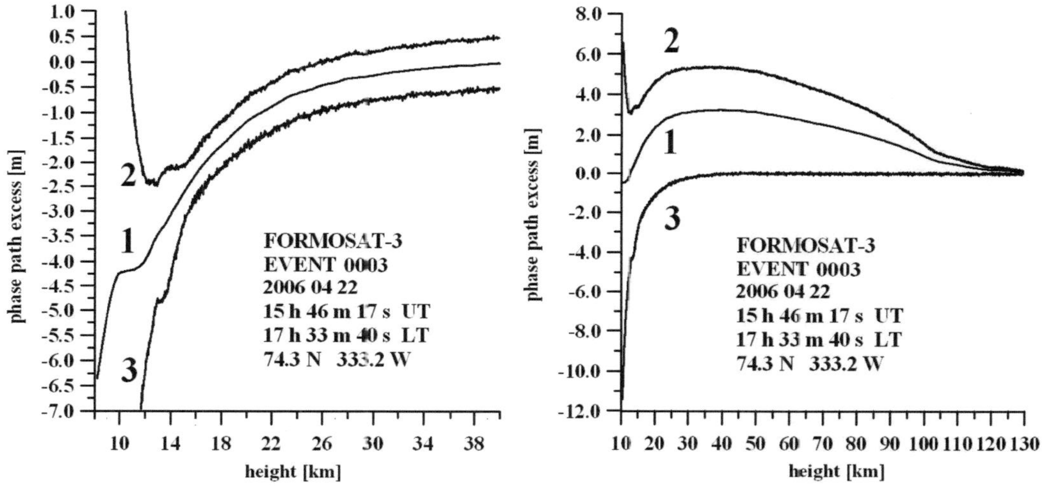

Fig. 4.2.2. The phase excesses $\Phi_1(t)$ and $\Phi_2(t)$ (curves 1 and 2) before (right) and after (left) ionospheric correction. Curves 3 describe the automatically corrected phase delay $\Phi(t)$ by use of Eq. (4.1.2). Curves 2 and 3 (left) are displaced by 0.5 m for better comparison. The noise, which is seen at frequency f_1 (curves 2), is transmitted after linear ionospheric correction to the contribution from the neutral atmosphere (curves 3). The ionospheric correction, which uses the ionospheric model, does not allow the noise transmission to the corrected phase delay at frequency f_1 (curve 1 in the left).

The model dependent ionospheric correction minimizes the noise level in the retrieved atmospheric refractivity. This allows obtaining high accuracy in vertical profiles of pressure and temperature. Results of retrieving vertical profiles of refractivity $N(h)$ [N -units], pressure $P(h)$ [hPa] and temperature $T(h)$ [K] from the phase delay $\Phi_1(t)$ are demonstrated in Fig. 4.2.3 for the FORMOSAT-3/COSMIC RO event № 0003, April 22, 2006, 17 h 33 m 40 s LT, with geographical coordinates 74.3 N 333.2 W. Data, shown in Fig. 4.2.3, left, are relevant to vertical profiles of pressure $P(h)$ (curve 1), refractivity $N_m(h)$, obtained from the standard model of the atmosphere (curve 2), refractivity $N(h)$, retrieved from the phase delay $\Phi_1(t)$ (curve 3), and to the difference $N_m(h) - N(h)$ (curve 4). The retrieved vertical profile of the refractivity is below the standard atmospheric refractivity profile because of geographical location in the North Polar region, where the cold temperature in the lower troposphere necessitates the higher rate of diminishing of refractivity and pressure with altitude. Vertical profile of temperature $T(h)$ obtained after the model-dependent ionospheric correction from the phase delay $\Phi_1(t)$ is shown in Fig. 4.2.3, right. The changes of temperature $T(h)$ are concentrated in the 216 - 228 K interval when the altitude is increasing from 10 km to 40 km.

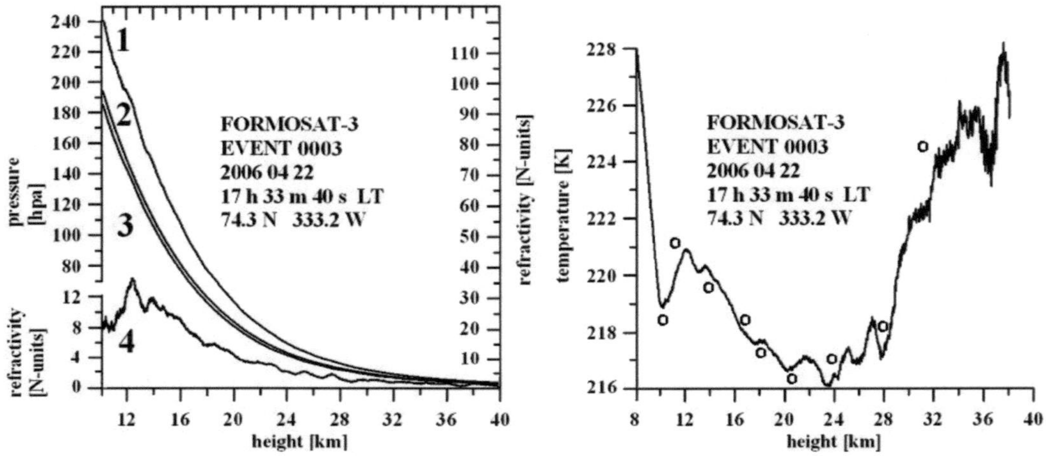

Fig. 4.2.3. Left: Results of retrieving vertical profiles of refractivity $N(h)$ and pressure $P(h)$ from the phase delay $\Phi_1(t)$. The top left vertical axis corresponds to pressure (curve 1) and the bottom left vertical axis is relevant to the refractivity perturbations (curve 4). The right vertical axis corresponds to altitude profiles of the refractivity calculated by use of standard atmospheric model (curve 2) and to the refractivity retrieved from the RO data (curve 3). Right: Results of retrieving vertical profile of temperature $T(h)$. Data are relevant to the west part of the Barents Sea with geographical coordinates 74.3 N 333.2 W. Open circles indicate the results of NCEP analysis for this area (April 22, 2006, 15 h 00 m UT http://www.cdc.noaa.gov/cgi-bin) at the altitudes corresponding to pressure 250 hPa; 200 hPa; 150 hPa; 100 hPa; 70 hPa; 50 hPa; 30 hPa; 20 hPa; 10 hPa.

The influence of tropopause, where vertical gradient of temperature changes from –5 K/km to +2 K/km is evident in the 8 km - 11 km altitude interval. The 11 K temperature increase with vertical gradient 0.9 K/km is visible between 28 and 40 km. The temperature obtained from RO measurements can be compared with data of NCEP analysis in this region relevant to the time of RO measurements. The results of NCEP analysis have been obtained from the website http://www.cdc.noaa.gov/Composites/hour/ and are shown by open circles at the altitudes corresponding to pressure values 250 hPa; 200 hPa; 150 hPa; 100 hPa; 70 hPa;

50 hPa; 30 hPa; 20 hPa and 10 hPa. Comparison of results of RO measurements analysis with NCEP data indicates an accuracy about of ±0.5° K between 12 and 20 km and from ±1 K to ±2 K between 20 and 35 km. This indicates, that quality of FORMOSAT-3/COSMIC RO measurements corresponds to that one observed in CHAMP experiments. Influence of layered wave-like structures with amplitudes of about 1 K– 3 K and vertical wavelength 1 - 3 km is clearly seen between 11 and 28 km.

Temperature vertical profiles in two equatorial and two moderate latitude areas are shown in Fig. 4.2.4 and Fig. 4.2.5. These RO events are located in the equatorial and tropical part of the South America, respectively. The moderate latitude's RO events are located in the South part of Indian Ocean (geographical coordinates 57.5 S and 295.1 W) and in the Northern part of the Pacific Ocean (geographical coordinates 55.5 N and 194.6 W), respectively (Fig. 4.2.5).

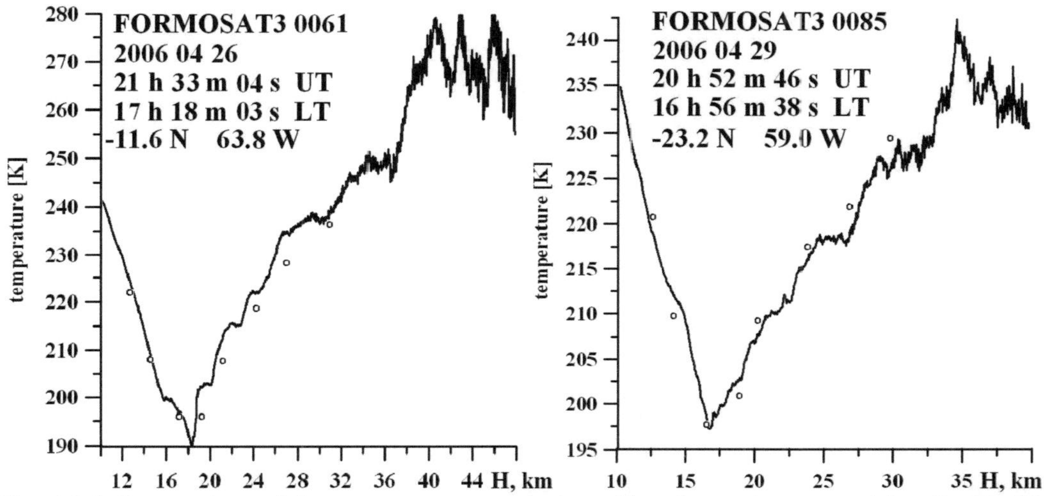

Fig. 4.2.4. Comparison of the temperature vertical profiles above the equatorial at 5 : 18 : 03 p.m. local time on Apr. 26, 2006, at 11°36′ N, 63°48′ W, and tropical part at 4 : 56 : 38 p.m. local time on Apr. 29, 2006, at 23°12′ N, 59° W of the South America (left and right panel, respectively).

The results of NCEP analysis are shown by open circles at the altitudes corresponding to pressure values 200 hPa; 150 hPa; 100 hPa; 70 hPa; 50 hPa; 30 hPa; 20 hPa; and 10 hPa. The forms of vertical temperature profiles are very different. In the equatorial and tropical regions of the South America the tropopause is clearly seen in the temperature profile with minimal temperatures between 190 and 197 K concentrated near the 17 km altitude. Above the tropopause the temperature is increasing from 190 - 196 K (Fig. 4.2.4, left and right) at h~17 km up to 270 - 275 K at altitude 46 km (Fig. 4.2.4, left) and up to 235 - 240 K at altitude 37 km (Fig. 4.2.4, right). Temperature variations observed between 17 and 40 km may be connected with influence of the internal atmospheric waves. Above the 36 km (Fig. 4.2.4, left) and the 32 km (Fig. 4.2.4, right) the temperature achieved maximal values ~ 270 K and 240 K, respectively. These maxima may correspond to warming effect of the stratospheric ozone. Random temperature variations between 36 km and 46 km altitude interval are mainly caused by the high-frequency receiver noise. In the moderate latitude area one can see the multiple tropopause phenomenon (Fig. 4.2.5). The first temperature minimum is located near the 10 km altitude. The minimal temperatures are also seen between 10 km and 22 km and in 10 - 28 km height intervals (Fig. 4.2.5, right and left, respectively). Small temperature variations between 10 km and 28 km height interval are connected with influence of the internal waves having vertical wavelength 1 - 4 km. The temperature increases by 30–40 K at the altitudes 22 - 40 km up to maximal value ~ 250 K–255 K at the 36 and 40 km (Fig. 4.2.5, right and left). Statistical error in the temperature measurement is quickly increasing and is about ±4 K at the altitude 40–46 km (Fig. 4.2.5). Comparison of results of RO analysis with NCEP data indicates good quality of FORMOSAT-3/COSMIC RO measurements.

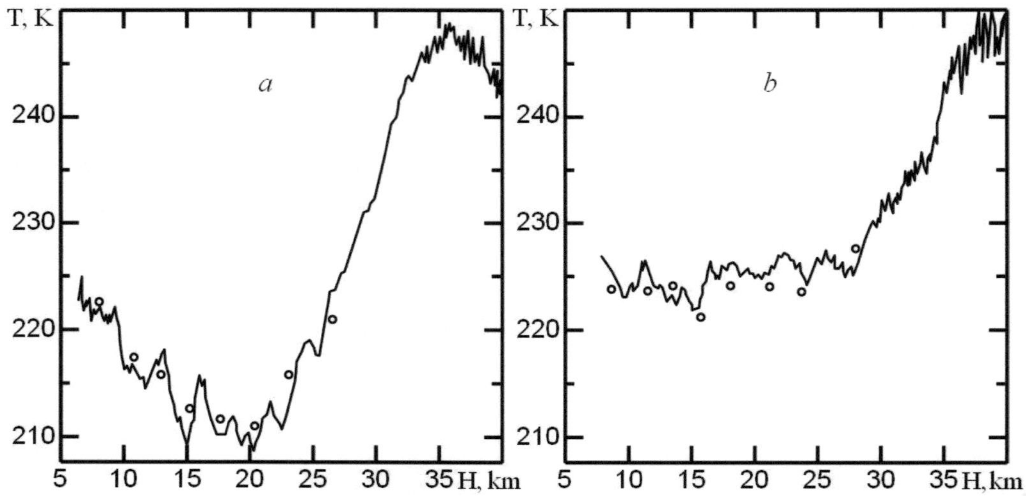

Fig. 4.2.5. Comparison of the temperature vertical profiles in the moderate latitude areas above the Southern part of the Indian Ocean at 4 : 55 : 01 a.m. local time on Apr. 23, 2006, at 57°48′ N, 295°6′ W, and in the Northern part of the Pacific Ocean at 3 : 05 : 54 p.m. local time on Apr. 25, 2006 , at 55°30′ N, 194°36′ W (left and right panel, respectively).

Therefore, results shown in Figs. 4.2.3- 4.2.5 indicate sufficient level of accuracy of temperature vertical profiles ± 0.5- ± 2 K), found from the FORMOSAT-3/COSMIC RO phase $\Phi_1(t)$ data, and reveal advantages of application of RO method to study meteorological parameters at different altitudes in the stratosphere and upper troposphere.

4.3. Amplitude of RO signal as radio-hologram of layers in the atmosphere

The amplitude channels of RO signal open new potential and capability for the observing and researching the atmospheric layers [98, 99]. As shown in [65, 66, 73, 98] the amplitude channels of the radio holograms contain important information on vertical gradients of the temperature and refractivity in the atmosphere and the electron density in the mesosphere. Quasi-regular structures are often seen in the amplitude data at heights corresponding to the tropopause and lower stratosphere. These structures change with geographic position and have vertical periods in the range of 0.8–4 km. The structures with small vertical periods are possibly associated with the atmospheric internal waves transmitting energy and momentum from the troposphere to the stratosphere [100, 101]. In Fig. 4.3.1 and Fig. 4.3.2

the amplitude variations A1 and A2 at two frequencies $L1$ and $L2$ are compared with the retrieved perturbations in vertical gradient of the refractivity $dN(h)/dh$ (curves G1 and G2) for GPS/MET RO events. Fig. 4.3.1 and Fig. 4.3.2 describe variations of the magnitude $dN(h)/dh$ for a low latitude region with coordinates 16.3 N, 104.6 W; and a moderate latitude area with coordinates 48.1 S, 333.7 W. The amplitude data A1 and A2 are normalized relative to the free space conditions and displaced for comparison. Curves G1 and G2 which describe the perturbations of $dN(h)/dh$ measured at frequencies $L1$ and $L2$, respectively, are also displaced in Fig. 4.3.1 and Fig. 4.3.2 for comparison. Fig. 4.3.1 and Fig. 4.3.2 consist of two parts describing perturbations of $dN(h)/dh$ in the altitudes intervals 2–20 and 20–40 km, respectively, with a goal to show variations of the refractivity gradient at high altitudes in an appropriate scale. The retrieved vertical refractivity gradients G1 and G2 reveal quasi-regular variations at the tropopause altitudes both in the low latitude and moderate latitude areas. Typical values of these variations are about ±(4–6) N-units/km and typical vertical periods are 0.8–2.5 km. Changes in vertical gradients of the refractivity may be connected with internal waves propagating through the tropopause areas and, consequently, may correspond to changes in the horizontal wind velocity. The changes of $dN(h)/dh$ below tropopause altitudes may be connected with water vapor influence and clouds structures. In the stratosphere region the wave-like structures can be seen in Fig. 4.3.1

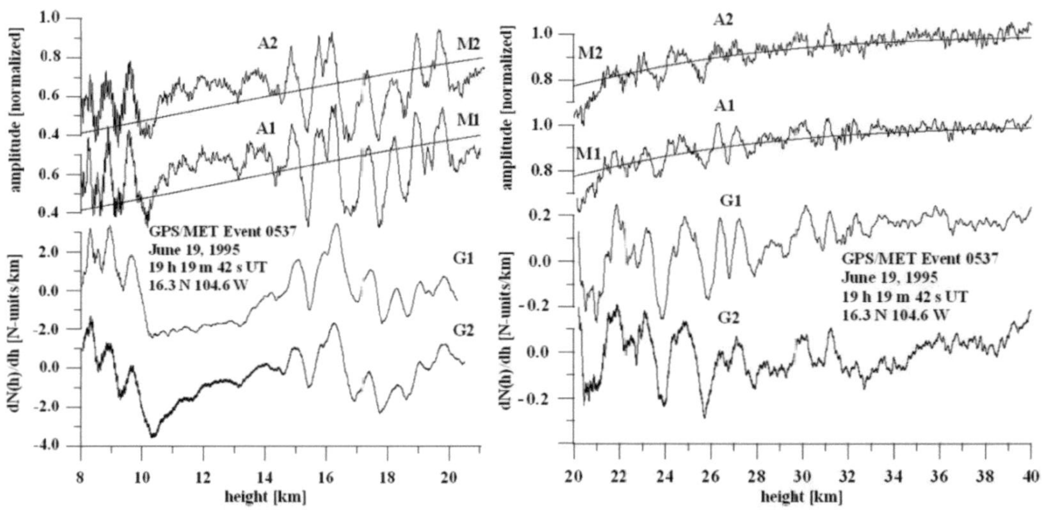

Fig. 4.3.1 Vertical gradients of the refractivity G1 and G2 (perturbed part) in the low latitude troposphere and stratosphere (above the Mexican coast of Pacific Ocean) retrieved from the amplitude channels A1 and A2 of radio holograms at frequencies F1 and F2 (GPS/MET radio occultation data, event 0537). Curves M1, M2 indicate the refractive attenuation corresponding to the refractivity model in the RO region.

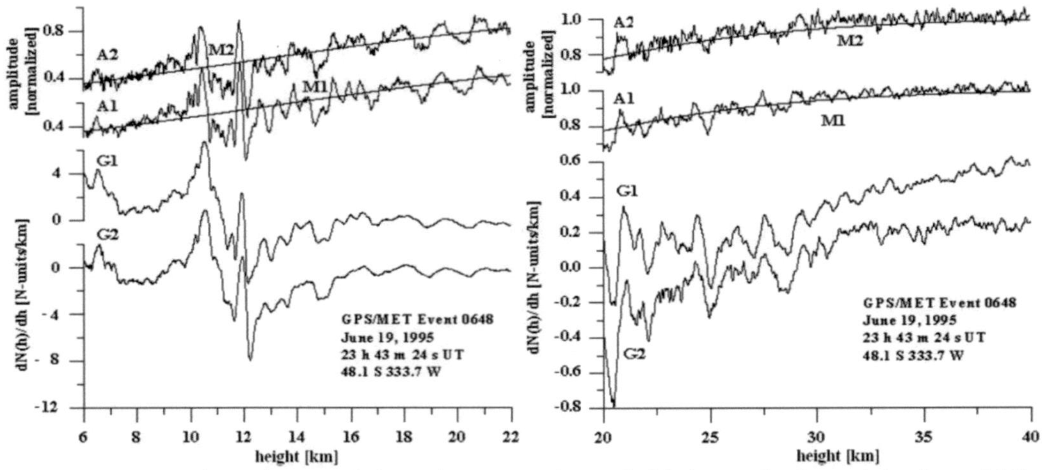

Fig. 4.3.2. Vertical gradients of the refractivity G1 and G2 (perturbed part) in the middle latitude troposphere and stratosphere (GPS/MET radio occultation data, event 0648). The designations are the same as in Fig. 4.3.1.

and Fig. 4.3.2 (right parts) also. Typical values of these variations are about ± 0.2–0.6 N-units/km and typical vertical periods are 1–3 km. Vertical gradients of the refractivity can be applied to retrieve vertical distribution of the temperature gradient $dT(h)/dh$. The restored vertical profiles of $dT(h)/dh$ are shown in Fig. 4.3.3 for the considered GPS/MET RO events. Curves GU in Fig. 4.3.3 present results obtained by the UCAR (University Consortium for Atmospheric Research) science team [32, 38] and published at the Internet site: http://www.cosmic.ucar.edu/gpsmet/feedback.html. Curves GU in Fig. 4.3.3 are obtained from the UCAR temperature profile TU by numerical differentiating.

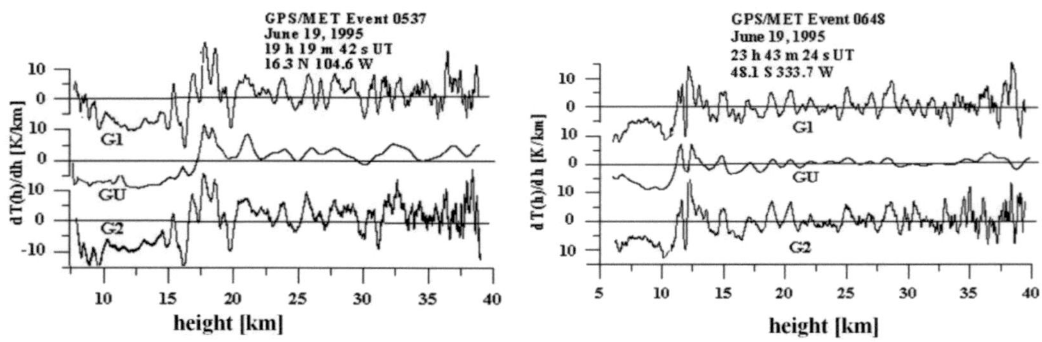

Fig. 4.3.3. Altitude profiles of temperature vertical gradient retrieved from the amplitude (curves GA) and phase RO data (curves GU) at two frequencies F1 and F2 for GPS/MET RO events 0537 (left) and 0648 (right).

In general the trends of the retrieved temperatures vertical gradient are consistent with those provided by UCAR. Temperature gradient retrieved from the amplitude data reveals better spatial variations with vertical periods 0.4–5 km than the UCAR temperature gradient. Sharp variations in vertical gradients corresponding to the tropopause region from –8 to –10 K/km up to 8 to 10 K/km are visible in Fig. 4.3.3. These changes are better seen in gradients restored from the amplitude data. Corresponding temperature variations have amplitudes ±(5 - 8) K in the tropopause and ±(2 - 4) K in the stratosphere. Variations of vertical gradient in the stratosphere reveal layered structures with vertical periods of about 0.8 - 2 km and amplitude of about ±(4 - 6) K/km. As follows from analysis of data shown in Figs. 4.3.1 - 4.3.3 the suggested method is useful for studying layered structures in the atmosphere.

4.4. Comparison of RO method with other satellites and ground based technique

In work [38] comparison of data, obtained during RO experiments in 1200 regions, with results of evaluation of vertical profiles of temperature and refractivity provided in the National Center for the Prediction of the environment state (NCEP) and the European Center of Middle-scale Weather Forecast (ECMWF) is conducted. Also comparison with the results of measurements of meteorological zonds, remote sensing observations by usage of radiometers in radio- and infrared ranges, oriented in nadir or in the earth's limb directions, and also with data of stellar occultation, has been provided. Each of these methods has own accuracy characteristics, vertical resolution and trustworthiness of the obtained data, and this requires bringing the results of measurements or evaluations to the same time instants and to the same coordinates of the investigated region. As the common parameters, characterizing the atmospheric state and determined from results of measurements or evaluations, the temperature T or the refractivity N at the same height or at the same pressure are selected. Before comparison of the temperature and refractivity, determined by different methods, let us briefly characterize these methods and their accuracy following to work [38]. The meteorological zond measurements under the "ideal" conditions and careful calibration ensure the instrumental temperature and relative humidity measurement errors are about 0.5 K and several percents, respectively, excluding the case of very low temperature or very high relative humidity [102]. Level of instrumental errors may be achieved only in the conditions of constancy of the environment state. By use of meteorological zonds measurements from the surface up to the height near 20–30 km may be conducted. With increasing of the altitude error in temperature determination is increasing and achieved 1 K near a 250 mbar pressure level (10.5 km altitude) and is heightening up to 4 K at 31 km altitude corresponding to pressure 10 mbar. Meteorological zonds measurements ensure the altitude resolution in the interval of several hundreds meters, however the temporal interval between launches of meteorological zonds occurs rare below 12 hours; some stations conduct meteorological zond's launches only one time per day and often in the different time. Meteorological zonds measurements give the most accurate information, used in centers of weather forecast, such as NCEP and ECMWF. These data play an important role for calibration and interpretation of results of radiometric limb measurements, conducted by a geostationary satellite of operative control of the earth's environment (GOES) and TV system for infrared observation (TIROS) of the operative vertical sounder (TOVS). Measurements by use of multi-channel radiometers, installed onboard GOES, ensure the temperature restoration at 40 pressure levels up to the height $h \approx 70$ km with the pressure equal to 0.1 mbar, and the humidity at 20 levels up to height $h \approx 17$ km with the pressure equal to 100 mbar in the case of the cloudless atmosphere. Advantage of this observation system is a possibility to obtain information about the altitude distribution of the water vapor. TOVS ensures temperature and pressure restoration with accounting for the humidity from results of measurements, conducted by three radiometers [103]. The stratospheric temperature and pressure data are obtained also by satellite observations of the oxygen emission in direction of the earth's limb by use of 15-channel radiometer at frequency 63 GHz. These measurements ensure obtaining the temperature data in the altitude range approximately from 25 up to 55 km where the pressure is changing from 46 up to 0.46 mbar. Accuracy of temperature determination is changing from 2.1 K up to 4.8 K at 22 mbar and 0.46 mbar pressure levels, respectively. At the 46 mbar pressure level error in temperature determination increases up to 7 K. Information on the altitude profiles of temperature in the altitude interval from 35 up to 80 km are obtained from results of stellar occultations. These measurements have vertical resolution 1.6 km and give temperature information at eight pressure levels in the interval from 4.6 up to 0.3 mbar with accuracy 2.5 K.

In [38] a technique for comparing results of determination of the refractivity and temperature from measurements provided by different methods were elaborated. The results of evaluations from NCEP and ECMWF models have been compared with data of meteorological zonds and measurements provided by different satellite missions including the RO observations. Values of refractivity and "dry temperature" have been compared at the same heights. The dry temperature is below the atmospheric temperature, however, it is restored immidiately from RO measurements and therefore it is convenient to use for comparison with results, obtained from different methods. Note that results of the temperature evaluations fulfilled by use of the NCEP and ECMWF models are discerning not greater than by 1 K with RMS variations from 1 up to 2 K. The refractivity values estimated from these models agree in the interval 0.5% at the altitudes from 2 up to 24 km with the RMS deviation below 1% above 10 km; in the lower troposphere the RMS deviation of values N is increasing up to 4%. Comparison of the restored from RO data vertical profiles of the refractivity $N(h)$ and temperature $T(h)$ and profiles $N_p(h)$ and $T_p(h)$, evaluated from the NCEP and ECMWF models, fulfilled in [38], showed, that the average difference $\Delta N(h) = N(h) - N_p(h)$ does not exceed 1% in the altitude range from 1 up to 50 km. In the altitude interval from 1 up to 30 km the RMS deviation of the difference in values of the refractivity $\delta(\Delta N)$ measured by RO method and evaluated from the NCEP and ECMWF models is located in the interval 1–2%. The average difference between the measured and evaluated temperature $\Delta T(h) = T(h) - T_p(h)$ is equal to 1 K in the altitude interval 1 km - 42 km. The RMS deviation $\delta(\Delta T)$ of the difference between the dry and actual temperatures found from measurements and evaluations is equal to 2 K in the altitudes interval 10 km - 25 km and increases below 10 km and above 25 km. Increase of $\delta(\Delta T)$ below 10 km is caused by significant variations of the water vapor content.

It was established from comparison of the RO and radiozonds measurements of the refractivity N and temperature T, that values N are coincided in the interval 1 - 2% in the altitude interval 2 km - 25 km. The temperature difference ΔT in average does not exceed 0.5 K in the altitude interval from 2 up to 28 km with the RMS values of $\delta(\Delta T)$ from 2 K - 3 K above 10 km. Below 10 km values of $\delta(\Delta T)$ are changing in more broad interval because of variations in the water vapor content. Comparison of RO data, results of the radiometric and other methods of N and T determination showed their good correspondence. RO measurements and data of limb microwaves sounder (MLS) are in very good agreement in average with values of N and T at the altitudes 24 - 45 km. The average difference in temperature ΔT does not surpass 1 - 2 K. 32 vertical profiles of $N(h)$ and $T(h)$, obtained by RO method, were used for comparison with radiometer data of the satellite *GOES*. The difference in values ΔN, determined by two methods, was about of 1% in the height interval from 4 up to 20 km with the RMS deviations $\delta(\Delta N)$ in the intervals 1 - 2% and 2 - 3% at the altitudes 10 - 20 km and 4 - 10 km, respectively. Values N, determined from RO data and from TOVS measurements, in average are coincided in the interval 1% at the altitudes 10 - 44 km with the RMS variations $\delta(\Delta N) = 1$ - 2 % in the altitude interval 10 - 30 km. Deviations of the average temperatures at the altitudes 10 - 40 km is of order 1 K with the RMS value $\delta(\Delta T) = 2$ K. According to TOVS data the temperature above 40 km is by 9 K lower, than from RO measurements, this may be because of the low resolution of TOVS sounder in a region of temperature maximum in the stratopause. Comparison of the radio and optical (HALOE) occultation measurements demonstrates coincidence of values N in the interval 1 - 2 % in the altitude interval 35 - 45 km with the RMS deviation $\delta(\Delta N)$ of about 2 - 3 %, correspondence of temperature in the interval 1 - 2 K was observed up to 40 km. Conducted in [104] comparison of nearly 170,000 profiles $N(h)$, obtained by satellite CHAMP, with the ECMWF forecast analysis showed nearly full coincidence of values N in the altitude interval 10–30 km with the RMS deviation of about 1%. Values of the RMS deviation depend on the geographical latitude. In the tropical regions they are equal to 3%. The RMS deviation of $N(h)$ is below 1% near the earth's surface in the moderate latitudes and in the polar region.

On the basis of the conducted in [38, 93, 104, 105] verification of RO results with other independent measurements one may conclude, that values of the refractivity in general case are coinciding with accuracy 1 - 2 %, and dry temperature - in the interval 1 - 2 K. In measurements, conducted by different methods, values of the refractivity in general demonstrated better agreement, than the corresponding values of temperature, since the temperature in the lower troposphere is significantly influenced by spatial variations of humidity. Therefore, the accuracy of RO method in determination of vertical profiles of atmospheric parameters is sufficient for use in the short-scale and middle-scale weather forecasts in the NCEP and ECMWF models.

Sufficiently large difference in the measured by different methods vertical profiles $N(h)$ is observed in the tropical regions. In this connection it is interesting to compare the results, obtained by RO and other methods in the regions with heightened concentration of water vapor in the atmosphere. In work [105] a comparison of the 33 vertical profiles of temperature, obtained during the RO GPS–MICROLAB experiments in the tropical region, was conducted with the radiozonde data in areas, separated by several hundred kilometers. The meteorological stations were disposed in the Indonesia in the regions with the geographical coordinates 6.9° S, 107.6° E (Bondung) and 0.03° N, 109.3° E (Pontianak). For comparison the data were used with difference in time not more than ± 6 hours. The RMS deviation of the temperature difference $\delta(\Delta T)$ from meteorological and RO measurements was about 1 K in the upper troposphere, where influence of humidity is small, and nearly 2 K in the stratosphere below 25 km. In the troposphere below 9 km the systematic distinction of the temperatures determined by two methods was observed. This distinction is caused by influence of humidity. The authors of work [105] showed that in the investigated regions the humidity influence appears already at the altitudes 8 - 10 km. Therefore, during analysis of results of RO sounding in the tropical region it is necessary to account for influence water vapor at the altitudes below 10 km. In work [38] it was shown that after accounting for influence of water vapor the RO and meteorological temperature measurements in the troposphere are coincided with accuracy about of 1 K.

Data of RO experiments in the moderate latitudes have been compared with the radiozonde measurements provided in the regions with coordinates 34.9° N, 140.09° E (Tsukuba, Japan). Comparison showed that the RMS values of the temperature difference $\delta(\Delta T)$, obtained by two methods, does not exceed 1.9 K in at the altitudes 14 - 30 km and are diminishing up to 0.9 K at the altitudes 5 - 14 km. At the altitudes below 4 km the RMS values of the temperature difference are in the neighbourhod of 3 K. This is caused by influence of the water vapor on the results of RO measurements.

Comparison of the vertical temperature profiles, obtained by use of the earth-based Rayleigh's lidars, with 18 RO temperature profiles, showed that the RMS values of the temperature difference is equal to 10 K in the altitude interval 40 - 60 km and are diminishing up to 6 K in the altitude interval 30 - 40 km. The authors of work [105] supposed that the relatively large distinction in these temperature measurements appears because RO soundings of the atmosphere were conducted above regions, separated by the distances up to 1000 km from a lidar. From comparison of the RO and lidar measurements having the time difference about one day, and spaced apart at 310 km, the values $\delta(\Delta T)$ were at around 4.3 K at the altitudes 40 – 60 km and nearly 2 K below 40 km height.

The tropopause plays significant role in the atmospheric processes determining the weather and climate. The tropopause is the transfer region between the troposphere which state is mainly determined by dynamic processes, and stratosphere, in which the emission and absorption processes prevail. The authors of work [106] found high correlation between the measured and evaluated pressures in the tropopause and reveal the seasonal and diurnal variations in the pressure fields on the basis of comparison of the high-resolution RO measurements and evaluations fulfilled in the National Center for the Prediction of the state of environment (NCEP) and in the National Center of the Atmospheric Research (NCAR).

The tropical tropopause plays the key role in connection between the tropical region and the middle atmosphere. The seasonal variations of the minimal temperature T_t and height h_t of the tropopause were investigated by use of the MICROLAB RO experimental data obtained near Bandung (Indonesia) at the latitudes 5° N - 15° S and longitudes 90° E - 125° E [105]. Comparison of the radiozonde and RO measurements showed, that the observed seasonal variations of T_t and h_t are corresponding sufficiently. The temperature difference does not exceed 2 K, and the RMS deviation of the minimal altitude difference, determined by two methods, is about 0.7 km. Important result of comparison of these two methods is demonstration of the sufficient accuracy of RO measurements for detecting and determination of the seasonal and diurnal characteristics of the temperature variations structure near the tropopause. Analysis of the tropical tropopause parameters obtained in period May 2001 November 2003 by use of the CHAMP satellite, and their comparison with the ECMWF evaluations results leads the authors of work [107] to the conclusion, that at the 30 - 300 hPa pressure levels the difference of the experimental and evaluated temperature values is well below 0.5 K. An annual circle and spatial variability of the tropical atmosphere and indicators of quasi-biannual variations have been revealed. The seasonal variations of the middle- and high- latitude tropopause were analyzed by use of 160000 CHAMP RO sessions data, conducted in the period May 2001 - July 2004. The results of RO determination of the "dry" temperature have been compared with analysis, fulfilled in NCEP and NCAR by use of the archives meteorological data [108]. Comparison of the RO and forecast data indicated their good correspondence and allows making conclusion on the effectivity of RO measurements for determination of the global distribution of the tropopause parameters. The accuracy of RO measurements of the atmospheric refractivity is around 1% at the altitudes 1 - 40 km. The errors in temperature determination do not exceed 1 - 2 K at the altitudes 2 - 35 km excluding the regions having high humidity. If the data on the altitude distribution of water vapor are known from independent sources the error in temperature restoration does not exceed 2 K. Therefore the RO sounding in the communication links satellite to satellite is an effective method of the meteorology and climate monitoring.

4.5. Investigation of layered structure and wave's activity in the atmosphere

The internal gravity waves in the atmosphere have been studied during many years. As it was turned out the small-scale waves with length about of several kilometers and period near 10 minute arise in the atmospheric boundary layer; waves with the horizontal length 100 - 1000 km, vertical period 1 - 10 km, and frequency 0.01 - 0.0001 s^{-1} are coming into existence owing to atmospheric fronts, cyclones, cumulus-nimbus convection, and also appear because of relief influence on the air movement in the mounting regions. Internal waves play an important role in vertical transfer of energy and kinetic momentum. Atmospheric circulation and temperature regime essentially depend on the intensity of internal waves, since they determine the turbulence development and mixing of the atmospheric layers. Waves breaking and dissipation play an important role in the dynamic and energetic balance of the mesosphere [109–112]. An origin of the internal waves is studied by different methods, including measurements of the middle-scale variations of temperature and wind velocity in the atmosphere with usage of the ground based radars. Radar measurements revealed a seasonal and annual cyclic recurrence in the internal wave activity in the moderate latitudes, connected with variations of location and intensity of the wind jets in the planetary frontal areas [113]. Airborne measurements indicated, that in the tropical zone the internal waves are arising, mainly, when an intensive development of the powerful cumulus-nimbus and thunder clouds takes place, particularly, in the areas of tropical depressions [114]. The seasonal and geographical distribution of stratospheric internal wave activity has been studied with usage of the meteorological rockets [115, 116]. For determination of the seasonal and latitudinal dependences of the potential energy of the internal waves the results of regular temperature measurements above Australia and Antarctica have been analyzed in the latitudes from 12° up to 68° S with usage of high-precision meteorological zondes. At the 17- 24 km altitudes in the stratosphere the annual variations of wave activity were revealed with maximum locating in the lower latitudes during the seasonal rains in December–January, and in the moderate latitudes - in the winter months (June–August) [117] . Measurements of wave parameters on rockets and with usage of radiozonde, radar and lidar observations are limited by positions of the ground based stations, located, mainly, in the Northern hemisphere [113–118]. For revealing of the mechanisms of the stratospheric and mesospheric dynamic it is necessary to elaborate models, including experimental information, distributed uniformly on the earth's surface, and describing an influence of waves on global variations of the air flows in the atmosphere. Satellite radiometric measurements allow obtaining information about dynamical structure of the internal waves in planetary scale [119, 120]. In particular, with usage of the satellite UARS the temperature variations with horizontal scales about 100 km were investigated and the distribution of internal wave activity in the atmosphere at 30 - 80 km has been found [119]. However, the ground based and satellite measurements were limited in time and have, mainly, a local character.

RO method lets new possibilities for studying of layered structures and atmospheric waves in global scale. A feature of internal atmospheric waves consists in smallness of vertical wavelength as compared with the horizontal one. With accounting for this feature the RO method is well adapted for observations of vertical structure of internal waves, since its vertical resolution is many times greater the horizontal resolution. One may detect the internal waves with horizontal size greater the horizontal resolution of RO method, which, as indicated in [121], may change from 100 up to 300 km depending on vertical period of waves. RO sounding of the atmosphere gives a possibility to study global wave activity with high vertical resolution and accuracy. The first analysis of possibilities of RO method for obtaining characteristics of the internal waves was conducted with usage of the satellite MICROLAB-1 data [100, 101, 122]. For each altitude temperature profile obtained from measurements of the phase path excess of radio waves, the dispersion of the temperature variations with vertical period from 2 up to 10 km has been determined, and the Brent-Vaisala frequency ω_b (i.e. the frequency of the free vertical air oscillations) has been found from the averaged temperature profile. By use of known theoretical relationships valid for the internal atmospheric waves from dispersion of the temperature variations and known values of the Brent-Vaisala frequency the seasonal and latitudinal variations of the potential energy of the atmospheric waves in the altitude interval 15 - 45 km in period 1995- 1997 years were determined with a global coverage [101, 122]. Unlike papers [100, 101, 122], that only the data of the phase measurements applied for investigation of the internal waves, in articles [55, 73, 82, 121, 123–125] a possibility for revealing characteristics of layered structures by use of joint analysis of the phase and amplitude variations of RO signal was investigated. It was indicated, that analysis of the altitude variations of the signal amplitude gives an additional possibility for investigation of the atmospheric waves. A technique for determination of vertical gradients of the refractive index and temperature in the atmosphere having high accuracy and spatial resolution was proposed and tested.

Let us consider a method and results of determination of the internal wave characteristics by RO technique. Amplitude and phase of signal react by different ways on the influence of layered structures and internal waves in the atmosphere. Quantitative estimation of the influence of layered structures on the amplitude and phase of radio waves may be obtained from expressions for the phase path and refractive attenuation of the radio waves. In papers [55, 82, 125] an important connection between the eikonal acceleration, Doppler shift, caused by the atmosphere, and intensity variations of the RO signal was grounded theoretically and revealed from an analysis of the radio-hologram, registered with usage of the satellites CHAMP and FORMOSAT-3. According to these papers, the second derivative of the phase path excess Φ is estimated from the relationships (3.4.15). This connection gives a possibility to recalculate the eikonal acceleration (or the derivative of the Doppler frequency with respect to time) into the refractive attenuation. This connection is important also for revealing and determination of parameters of layered and wave structures by means of joint analysis of information, contained in the amplitude and phase channels of the radio-holograms, since, in accordance with Eqs. (3.4.15), the contributions of layers in variations of the eikonal acceleration and intensity are correlated. Turbulent and irregular structures in the atmosphere give, as a rule, uncorrelated contributions in variations of intensity and eikonal acceleration. For revealing of influence of atmospheric layers on the phase path it is necessary to apply a method of analysis and data handling proposed in papers [82, 125]. Typical examples of variations of the amplitude of RO signal *E*, normalized relative to the level of free propagation, are shown in Fig. 4.5.1. The first curve corresponds to the atmospheric sounding in the Northern part of the Pacific tropical zone, the second is relevant to the Sachara desert. The third and forth curves characterize atmospheric influence on RO signal in the Northern and Southern polar areas. The fifth and sixth curves were obtained during measurements in the moderate latitudes of the Northern hemisphere. The wave structures having different intensity are seen in Fig. 4.5.1.

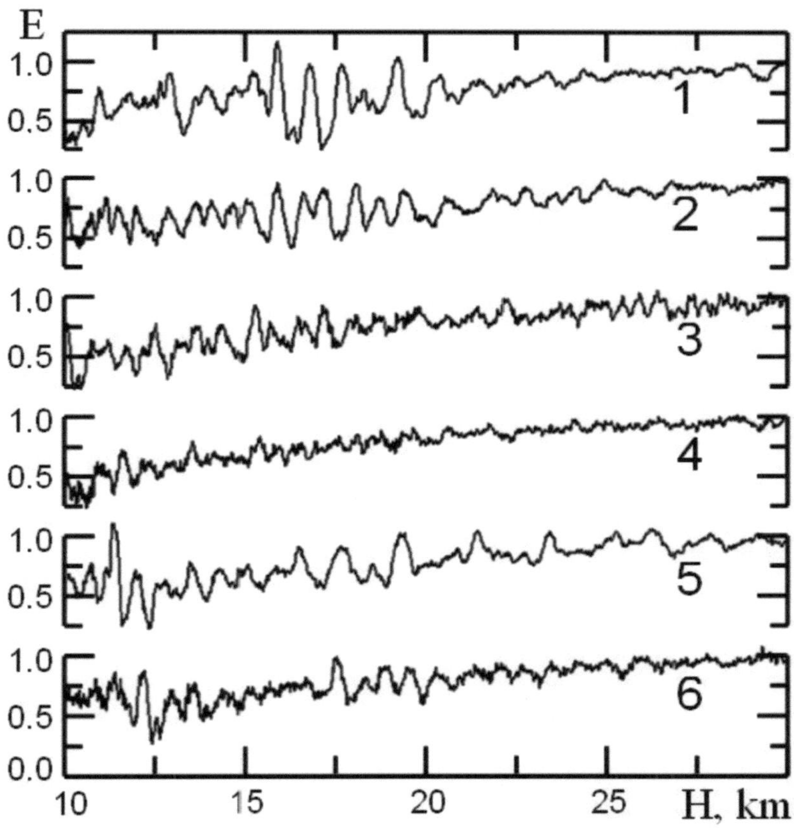

Fig. 4.5.1. Typical examples of vertical profiles of the amplitude of RO signal *E*, normalized relative to the level of free propagation.

The amplitude variations in the tropical atmosphere (curves 1 and 2) are most essential. In the polar zone the intensity of the amplitude variations are smaller, however, they occupy greater altitude interval (curves 3 and 4). Amplitude variations in the moderate latitudes occupy intermediate position relative to the tropical and polar regions (curves 5 and 6). It is indicated in papers [73, 123–125] how to obtain information on the altitude profile and vertical period of layered and wave structures in the atmosphere by use of analysis of variations of the signal amplitude. Influence of quasi-regular structures become apparent often in the altitude dependence of the amplitude data, corresponding to the tropopause and lower stratosphere, where these structures have vertical period in the range 0.8 – 4 km and may be connected with waves, transferring the energy and kinetic momentum through the atmosphere. Joint analysis of the phase and amplitude variations allows more reliable identifying of the waves and layers in the atmosphere. Let us consider the results of measurements that corroborated the connection between the amplitude and phase variations. The examples of observation of the influence of layered structures in measurements of the altitude dependence of the refractive attenuation $X(t)$ and comparison with results of evaluation of the attenuation X_p from the eikonal acceleration are shown in Fig. 4.5.2. The data shown in Fig. 4.5.2 were obtained during measurements, conducted with usage the satellites FORMOSAT-3. For evaluation of the refractive attenuation X_p from the phase data the coefficient m in Eqs. (3.4.16) was determined from trajectory data which contain the coordinates and velocities of the GPS and FORMOSAT-3 satellites. The correspondence between values of the refractive attenuation X_p and X, obtained from the phase and amplitude data, may be seen in Fig. 4.5.2. Correspondence between the altitude dependences X and X_p seems to be good, if one accounts for only the slow changing component, describing the gradual increase of the refractive attenuation when the height H decreases. High-frequency variations of the refractive attenuation, found from the amplitude and phase data, qualitatively coincide. Amplitude variations of the refractive attenuation diminish when the height increase, indicating, therefore, the atmospheric origin of the layered or wave structures. According to Fig. 4.5.2 the vertical period of wave structures changes in the 1– 4 km interval that is typical for internal atmospheric gravity waves [100, 101]. Small difference in high-frequency variations may be connected with influence of the diffraction irregularities and turbulence effects, and also by receiver noise. It follows from analysis of data, similar to indicated in Fig. 4.5.2, that the relationships (4.5.1) are fulfilling up to heights about of 30 – 40 km. Therefore the eikonal acceleration is an informative parameter having the same importance for the experimental data analysis as the caused by the atmospheric influence the Doppler shift of RO signal. Comparison of variations of the refractive attenuation, found from the phase and amplitude data, allows reliable revealing the contribution of atmospheric layers from the background containing the interferences caused by turbulence and random irregularities. With usage of the elaborated method it is possible independent determining characteristics of turbulence and contributions of random irregularities. With usage of the phase data one may correct systematic errors in measurements of the amplitude, caused by the gain instabilities in the radio links satellite-to-satellite. A feature of application of the eikonal acceleration technique consists in a possibility to diminish the influence of a regular ionospheric trend on restoration of the atmospheric parameters by use of the second differentiation on time.

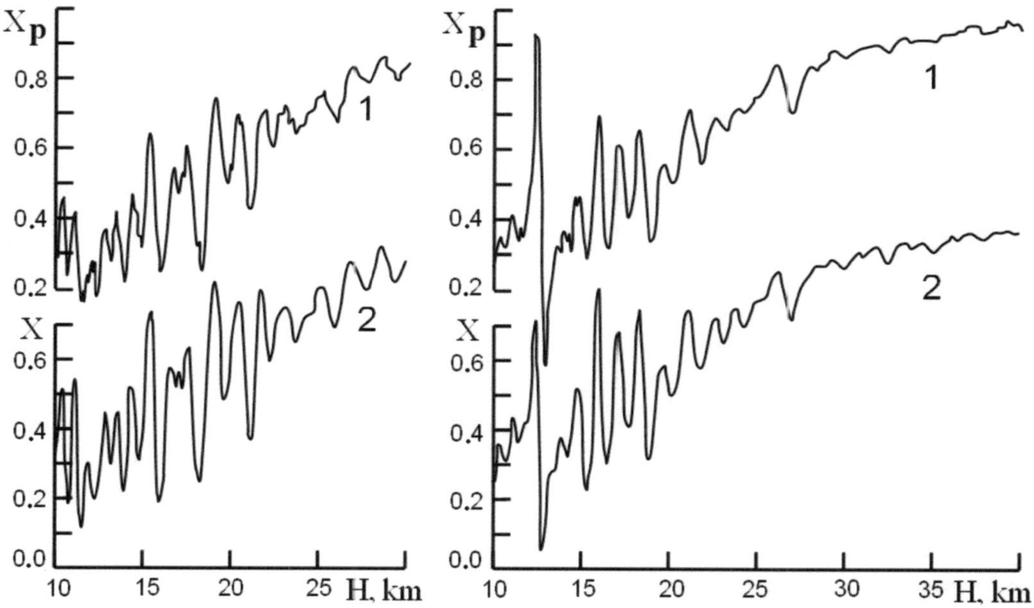

Fig. 4.5.2. The refractive attenuation, found from the phase and amplitude data.

Quasi-periodical layered structures, observed in vertical profile of the refractive index gradient, are connected with influence of internal atmospheric waves [100, 101, 122]. Internal atmospheric waves are characterizing by the amplitude variations of vertical gradients of the refractive index $a_N = |d(\delta N)/dh|$ or temperature $a_T = |d(\delta T)/dh|$, where δN and δT are the perturbations of refractive index and temperature relative to their background values in a given altitude interval. Let us consider in more detail method of restoration of the wave parameters from measurements of the amplitude variations of RO signals. For example, the amplitude variations indicated by curve 2 in Fig. 4.5.1 will be analyzed. These variations may be connected with atmospheric wave structures. Perturbations of vertical gradients of refractive index $d(\delta N)/dh$ were determined from the amplitude variations. Examples of variations of $d(\delta N)/dh$, restored from the amplitude data are indicated in Fig. 4.5.3. A wave structure is clear seen in changes of values $d(\delta N)/dh$ between the altitudes 8 and 40 km. The vertical period of wave structure increases from 0.8 - 1.0 km between 8 and 25 km up to 4 km between 30 and 40 km altitudes. Sharp change in $d(\delta N)/dh$ is observed at the height near 40 km, where breaking of internal wave occurs. From perturbations of vertical gradients of the refractive index one may determine variations of temperature and it gradients [125]. Examples of restored temperature variations and it gradients is shown in Fig. 4.5.4 (left). The vertical temperature gradient grows from ±2 K/km in the 8 - 20 km height interval up to ±8 K/km between 30 and 45 km altitudes, the temperature fluctuates from ±0.2 - ±1.0 K at altitudes 8 and 20 km up to ±4 K between 30 and 40 km heights. The vertical period of the temperature gradients $d(\delta T)/dh$ grows from 0.8 km between 8 and 20 km heights up to 4 km in the 35 - 40 km altitude interval. The most possible source of the observed structures with vertical period

0.8 - 4.0 km is an influence of the atmospheric wave [100, 101, 122]. If the observed structures are caused by the wave influence, then, as indicated in paper [116], the temperature variations may be connected with perturbations of the wind velocity with usage of the dispersion relationships, fulfilling for the internal gravity waves. The dispersion relationships connect vertical wavelength λ_v with the internal phase velocity of the atmospheric waves v_i, which may be measured by observer, moving with the velocity, equal to a background velocity U of the horizontal wind

$$\lambda_v = 2\pi v_i / \omega_b, \qquad v_i = |c_a - U\cos\varphi_a|, \tag{4.5.1}$$

where ω_b is the Brent-Vaisala frequency, c_a is the phase velocity of atmospheric wave, measured relative to the Earth, φ_a is the azimuth angle between the background wind velocity and direction of wave propagation [116]. Polarization relationships for waves, according to paper [110], connect the complex amplitude of temperature variations δT with perturbations of the horizontal wind velocity $v(h)$

$$v = \mathrm{Re}[ig(\delta T)(T_b\omega_b)^{-1}], \tag{4.5.2}$$

where $\omega_b = gT_b^{-1}\Gamma$, $\Gamma = dT_b / dh + 9.8^\circ$ K/km, g is the gravity acceleration and T_b is the background temperature at given height. The dispersion and polarization relationships foratmospheric waves may be applied when a condition that the wave frequency is greater, than the Coriolis parameter f_c, but many times smaller the Brent-Vaisala frequency ω_b. The Coriolis parameter f_c is connected with the angular velocity of the earth rotation Ω and latitude φ by a relationship $f_c = 2\Omega\sin\varphi$. From (4.5.2) one may obtain a connection between vertical gradients dv / dh and $d(\delta T) / dh$

$$dv(h)/dh = d\,\mathrm{Re}\left[ig\ (\delta T)\ (T_b\omega_b)^{-1})\right]/dh \approx \mathrm{Re}\ \left[ig\ (T_b\omega_b)^{-1}d(\delta T)/dh\right]. \tag{4.5.3}$$

Eqs. (4.5.3) is valid under assumption, that T_b and ω_b are slow changing functions of height at a scale about of vertical length of atmospheric wave. For determination of variations of the horizontal wind velocity one may apply the Hilbert transformation [126]. The Hilbert transformation gives an analytical presentation of function $d(\delta T) / dh$

$$d(\delta T)/dh = \mathrm{Re}\left\{a_T(h)\exp\left[i\ \Phi_T(h)\right]\right\}, \tag{4.5.4}$$

where real functions $a_T(h)$ and $\Phi_T(h)$ are the amplitude and phase of vertical

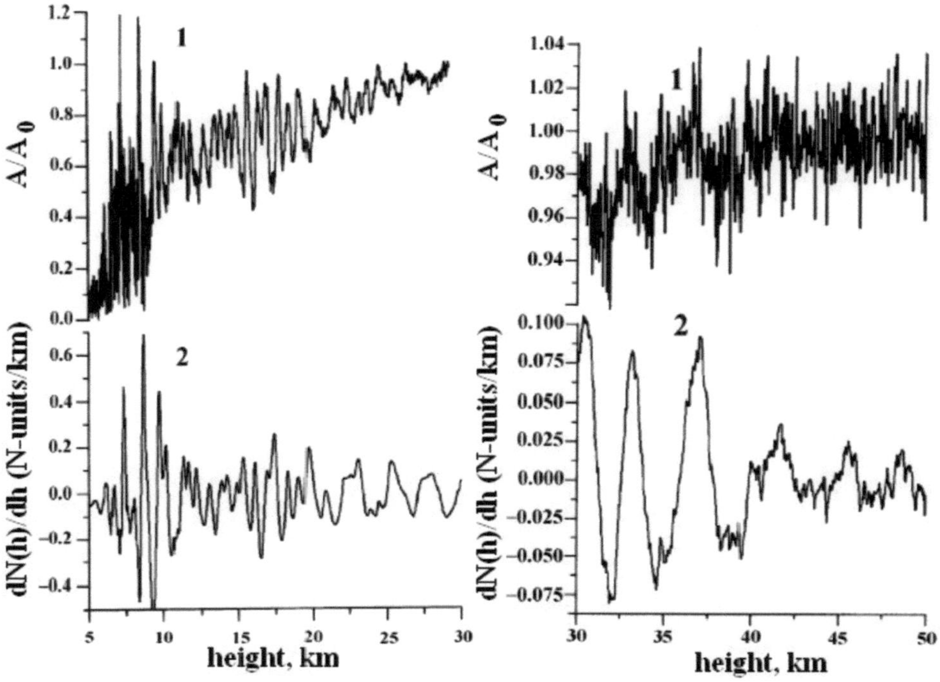

Fig. 4.5.3. Vertical profiles of $d(\delta N)/dh$, restored from the amplitude data.

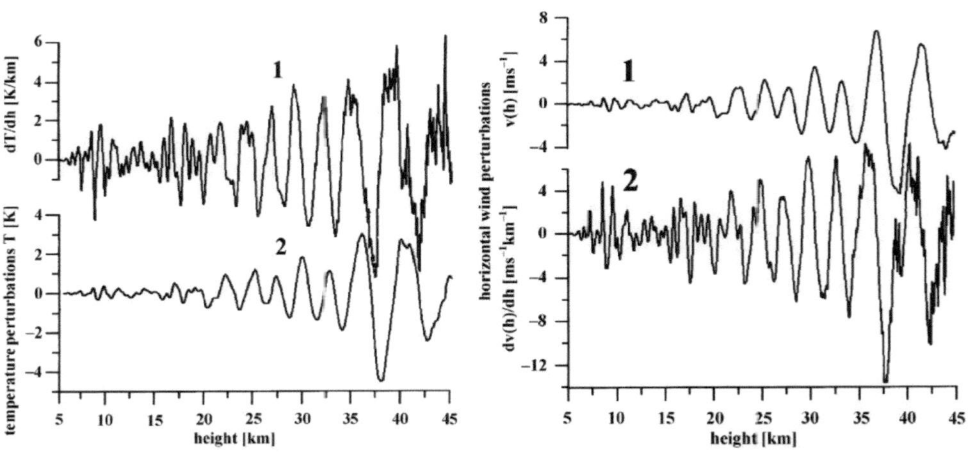

Fig. 4.5.4. Vertical profiles of the restored temperature variations and its gradients (left). Vertical profiles of the perturbation wind velocity (1) and its gradients (2) (right).

temperature gradients. The function $dv(h)/dh$ may be restored by use of Eq. (4.5.3) from variations $d(\delta T)/dh$. These relationships may be applied for estimation of the wind parameters. Results of restoration of vertical profiles of the perturbation wind velocity and it gradients are indicated in Fig. 4.5.4 (right). The amplitude perturbations of the wind velocity are indicated by curve 1, and change from ±0.5 m s^{-1} up to ±5.0 m s^{-1} when the height increases from 10 km up to 40 km. Corresponding value of vertical gradients of the perturbations wind velocity changes from ±2 ms^{-1}km^{-1} up to ±8 ms^{-1}km^{-1}, when the height increases from 10 km up to 40 km. An example of internal wave «portrait», which is described by vertical profile of the phase $\Phi_N(h)$ (curve 1) and amplitude a_N (curve 2), is

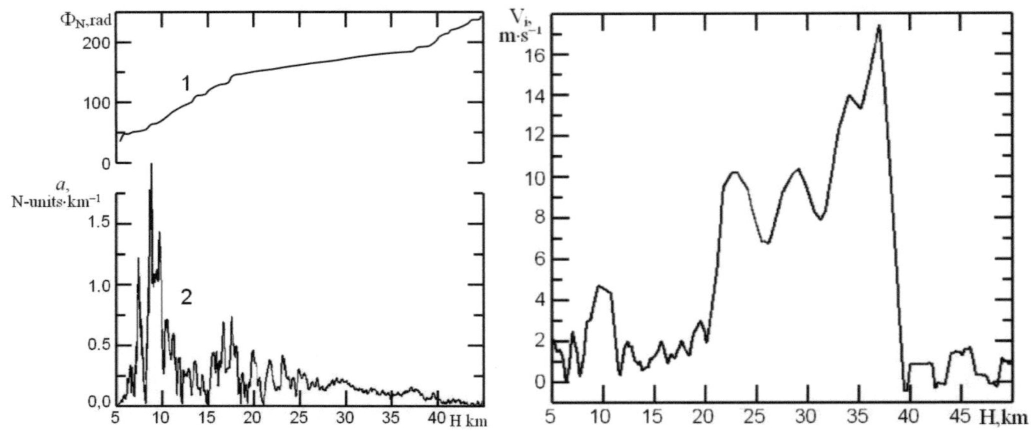

Fig. 4.5.5. Internal wave «portrait» consisting of the phase $\Phi_N(h)$ (curve 1) and amplitude a_N (curve 2) (left). Vertical profile of the internal phase velocity (right).

shown in Fig. 4.5.5 (left). It is necessary to note, that the wave «portrait» may be obtained from the amplitude variations of RO signals without any assumption concerning a nature or origin of the atmospheric wave. A sharp change of the phase of atmospheric wave is observed at the height, where the amplitude of internal wave diminishes below the noise level and the wave ceases it existence. At the height ~40 km a transfer of internal wave energy occurs in the energy of turbulent structures. The internal phase velocity of wave v_i may be estimated by means of differentiation of the wave phase $\Phi_N(h)$. Results of determination of the internal phase velocity are indicated in Fig. 4.5.5 (right). Value $v_i(h)$ changes in the interval 2 - 16 ms^{-1}. A sharp change of the internal phase velocity near the height 40 km is connected with influence of a boundary of wave breaking region.

The polarization and dispersion GW relationships can be used to find the horizontal wind perturbations and kinetic energy of GW as shown formerly in [124]. The GW contribution can be obtained after subtracting the regular trend from the refractive attenuations X and X_p . Curves 1 and 2 in Fig. 4.5.6, left top panel, demonstrate the refractive attenuations X and X_p evaluated from the eikonal acceleration and amplitude data for the FORMOSAT-3/COSMIC RO event 0003 (April 22, 2006, 17 h 33 m 40 s LT, geographical coordinates 74.3 N 333.2 W) at the first GPS frequency f_1. Smooth curves 3 describe the results of modeling of the refractive attenuation. The significant correspondence between the refractive attenuations obtained from the amplitude and phase data is clearly seen in Fig. 4.5.6, left top panel (curves 1 and 2). This correspondence is excellent when one consider the slow trend describing the diminishing of the refractive attenuation with height. The high frequency oscillations in the refractive attenuation found from the amplitude and phase data coincide quantitatively.

Previously in [66], [125] it has been showed, that the amplitude variations of RO signal can be used to find the vertical gradient of refractivity $dN(h)/dh$. Equation (3.4.15) opens a new way to find vertical gradient of refractivity by the same way as formerly used for the amplitude data [125]. Variations of vertical gradient of refractivity $dN_p(h)/dh$ and $dN_a(h)/dh$ [N -units/km] found from the FORMOSAT-3/COSMIC RO phase and amplitude data are compared in Fig. 4.5.6, right top panel (curves 1 and 2, respectively). There is good correlation between $dN_p(h)/dh$ and $dN_a(h)/dh$ up to 28 km. Typical values of vertical gradient variations are of about ±0.12 - 0.18 N-units/km, and typical vertical periods are 0.8 - 2.5 km. Changes in vertical gradients of the refractivity may be connected with the GW propagating through the tropopause areas and, consequently, may correspond to the changes in the horizontal wind velocity. At the 28- 40 km, the amplitude and form of wave structure are sharply changed. This indicates the wave-breaking zone in the 28- 30 km interval. Note, that existence of the wave breaking zone at latitudes 28–30 km in polar regions has been predicted earlier in publication [127]. The amplitude of wave structure diminished by 2- 3 times corresponds to energy dissipation by 4- 9 times. The wave-breaking effect and the instantaneous radio image of wave-breaking area are obtained by the FORMOSAT-3/COSMIC mission in the first time in RO practice simultaneously in the amplitude and phase data. This observation supported the preliminary conclusion, made earlier in [125], on a possibility to study the wave-breaking areas in the atmosphere by the amplitude RO method. If the observed wave structures are caused by the GW activity, then vertical gradients of refractivity can be related with horizontal wind perturbations using the GW polarization Eqs. [125]. The horizontal wind perturbations found from the phase and amplitude data are indicated in Fig. 4.5.6 (left bottom panel).

The altitude profile of the perturbations horizontal wind velocity, determined from variations of vertical gradients of refractive index with usage a method, described in [83], are demonstrated in the left bottom panel in Fig. 4.5.6. Curves 1 and 2 describe vertical profile of the perturbations horizontal wind velocity, found from amplitude and phase data, they correspond to vertical gradients refractive index, indicated in the right top panel in Fig. 4.5.6. Perturbations of the horizontal wind velocity v are located in interval $\pm$ 2 - 3) ms^{-1} between 28 and 40 km the amplitude and character of variations of the gradients of refractive index and wind velocity change sharply, that indicates existence of the internal waves breaking zone at this altitudes.

According to review [110], the perturbations of the horizontal wind velocity v may be recalculated in the kinetic energy of atmospheric waves E_k with usage a relationship

$$E_k = v^2/2 \quad , \tag{4.5.5}$$

where E_k has dimension J kg^{-1}. From Eq. (4.5.5) it follows, that decrease in the amplitude variations of the horizontal wind velocity by 2 - 3 times corresponds to dissipation energy by a factor 4 - 9. Potential wave energy may be found with usage of relationships

$$E_p = g^2(\delta T)^2(2T^2\omega_b^2)^{-1} = g^2(\delta N)^2(2N^2\omega_b^2)^{-1} \ , \tag{4.5.6}$$

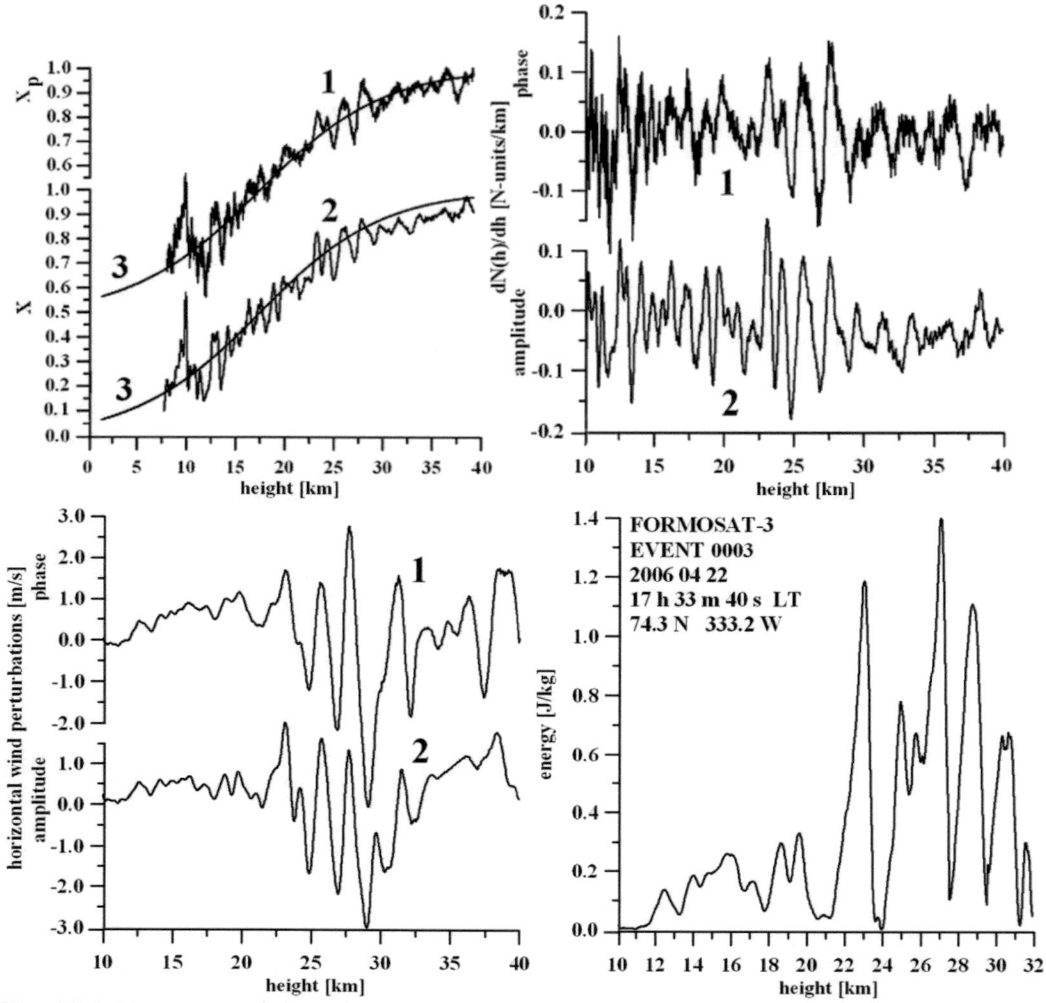

Fig. 4.5.6. Variations of the refractive attenuations retrieved from the amplitude and phase data. Variations of vertical gradient of refractivity (curve 1 and 2, right top panel). The altitude profiles of the horizontal wind perturbations (bottom, left panel). The sum of kinetic and potential energy of wave structure (bottom, right panel). Curve 1 and 2 correspond to physical parameters, retrieved from the amplitude and phase data. Curve 3 describes the refractive attenuation corresponding to the standard altitude dependence of refractivity.

where g is the gravity acceleration expressed in m/s^2, ω_b is the Brent-Vaisala frequency expressed in Hz, δT and δN are variations of the temperature and refractive index, $T(h)$ and $N(h)$ are the background altitude profiles of temperature and refractive index [100,101, 110, 122]. The energy of internal gravity wave is equal to the sum of kinetic and potential energy

$$E_p + E_k = g^2 (2\ \omega_b^{\ 2})^{-1} (\delta N)^2 N^{-2} + v^2 / 2\,. \tag{4.5.7}$$

Variations in the refractive index, δN, may be found by integrating vertical gradients, presented by curves 1 and 2 in the right top panel in Fig. 4.5.6. Results of energy determination are indicated in the right bottom panel in Fig. 4.5.6. The main part of the atmospheric wave energy is concentrated between 22 and 32 km, in this region occurs an active transfer process of the energy and kinetic momentum between different layers in the stratosphere. Interpretation of variations of RO signal parameters as a consequence of the atmospheric wave influence is grounded on results of comparison with measurements, accomplished by radiozondes [101, 122, 123].
RO method allows carrying out a global monitoring of atmospheric waves and finding the geographical and altitude distributions of their activity. It is necessary to note, that the amplitude of RO signal is sensitive to different types of internal gravity waves, and also to other waves, having small vertical periods. Activity of internal waves expressed as dependence on the height in the atmosphere will be characterized by the amplitude $q(h)$ of the Hilbert transformation of vertical gradients of refractive index. In Fig. 4.5.7 examples of the geographical maps showing the distributions of atmospheric waves intensity with the amplitude a_N, greater, than 0.6 N- units km^{-1}, are presented for period 15 July - 30 July 2003 year, and the heights 14 km (top) and 18 km (bottom). In [124] from results of RO measurements the detailed geographical distributions of the atmospheric wave's activity were obtained. The seasonal features were observed. It was indicated, that in the equatorial regions at the height 12 km waves activity is practically independent on the season. In the polar areas there is notable seasonal activity, which is maximal in summer. At the mid-latitudes of the Northern and Southern hemisphere waves activity has maxima at altitudes 14 km and 16 km. At the altitude 18 km the main energy of internal waves is concentrated near equator. In Siberia minimum of wave's activity at the height 14 km is achieved in winter, and maximum - in summer. In summer 2003 in the Northern Atlantic weak wave's activity was observed at the altitude 14 km. This may be caused by low level of meteorological activity in the troposphere. On the contrary, at the same altitude in the Northern-eastern part of Asia continent and in the Northern America strong wave's activity was observed. Wave's activity between 20 and 26 km altitude interval is concentrated, mainly, in equatorial zone. The intensity of internal waves in the equatorial region quickly diminishes when the height increases, that is connected with decreasing of air density. The longitude dependence of the internal wave intensity practically is absent. Exceptions are the maxima of activity in the regions of Northern America and Persian Gulf at the altitudes 20 - 22 km. It follows, that analysis of the amplitude variations of RO signal, passed through the atmosphere in the communication link satellite-to-satellite, is effective method of detecting and studying of wave activity on a global scale. The amplitude data have heightened sensitivity to wave's structures with small vertical periods, because the amplitude of RO signal is proportional to the square of vertical period of internal waves [83]. Proposed analytical method, based on the Hilbert transformation, allows finding vertical distribution of the amplitude and phase of atmospheric waves from results of measurements of the amplitude of RO signal, i.e. obtaining "image" of their vertical structure. In addition, in the case of internal gravity waves the dispersion and polarization relationships allows determining the internal phase velocity and vertical profile of the horizontal wind perturbations. The amplitude measurements allows obtaining the geographical distributions and seasonal dependence of global atmospheric waves activity. For example, analysis of RO data, obtained during 2003 year by satellite CHAMP, revealed an asymmetry in the

atmospheric waves activity . Maximum in waves activity was observed in summer in polar zone at the altitude 12 km. At the mid latitudes of the Northern and Southern hemispheres waves activity is maximal in the 14 - 16 km height interval. In the equatorial regions the main intensity of internal waves become apparent at altitudes 18 - 26 km. In some regions, for example, in Siberia, seasonal dependence of waves activity was detected: in winter 2003 at the altitude 14 km a low wave activity was observed, and in summer 2003 wave activity was high.

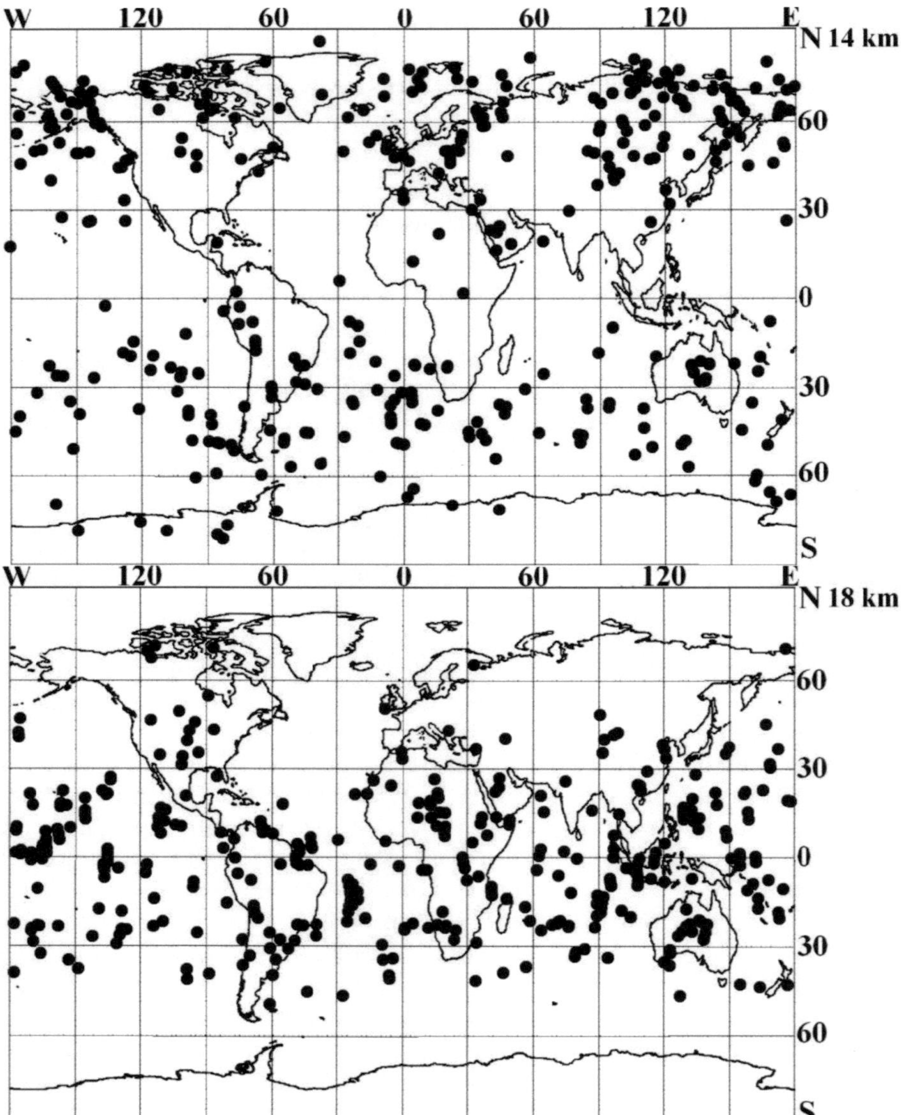

Fig. 4.5.7. Geographical distributions of atmospheric waves intensity with the amplitude a_N, greater, than 0.6 N- units km^{-1}

4.6 . Measurements of radio wave propagation effects: refraction, absorption, and diffraction

Effects of radio waves propagation are important for global real-time monitoring of the troposphere, stratosphere, and ionosphere, and for estimating conditions for telecommunication in trans-atmospheric satellite-to-satellite links. The atmospheric influence introduces the phase delay (or the phase path excess relative to the phase path in free space) and Doppler shift in the carrier frequency of the radio waves. Certain physical phenomena can attenuate the intensity of radio waves by refractive spreading (the refractive attenuation effect) and atmospheric absorption losses (the total absorption effect) as they propagate through the atmosphere. The atmospheric absorption losses at GPS frequencies are caused mainly by atmospheric oxygen [128]. The radio occultation technique is an effective tool for the investigation of radio wave propagation effects in the trans-atmospheric links and for remote sensing of the earth's atmosphere and ionosphere at different altitudes with global coverage [16, 37, 48, 52]. The general relations for changes in Doppler shift, phase, amplitude, bending angle, and absorption have been developed in [22]. Significant RO investigations of the radio wave's propagation effects began in Russia in 1990 with the use of the MIR orbital station and two geostationary satellites [29–31]. Radio links of the Ku ($\lambda = 2$ cm) and the UHF radio band ($\lambda = 32$ cm) with transmitters of increased power and antennas with high directivity were used. For many years analysis of the Doppler shift variations remained the only one practical way to obtain vertical profiles of the bending angle, impact parameters, and refractivity by use of satellites trajectory data. In recent years new radio holographic techniques based on combination of the RO amplitude and phase data have been considered, such as Fresnel diffraction theory [129–132], synthetic aperture method [69, 133, 134], radio-holographic focused synthetic aperture (RHFSA) [65, 70, 73, 74, 98, 135], canonical transform [79], and the Full Spectrum Inversion (FSI) method [75]. These studies improved vertical resolution and accuracy in retrieving of physical parameters of the atmosphere and broadened the applicability domain of RO method. This method may be applied also to identify the contribution of layers in the atmosphere and ionosphere and to estimate the total atmospheric absorption. The absorption measurements are planning for future RO missions [87] to determine with high vertical resolution the water vapor abundance at different altitudes in the stratosphere and troposphere. A differential Canonical Transform/Full Spectrum Inversion (CT/FSI) technique was proposed to retrieve absorption in X/K band, 9–22 GHz [88]. At decimeter wavelength band the total absorption effect in the trans-atmospheric telecommunication link orbital station MIR- geostationary satellites was measured at frequency 930 MHz [31]. In this experiment the refractive attenuation has been excluded by use of the phase and Doppler frequency data. A possibility to separate the absorption term and to measure absorption in the atmosphere in X/K band by use of a spectral phase matching method (SPMM) and Fourier Integral Operators (FIO) has been detected in [76, 77]. The authors of [76, 77] demonstrated how the second derivative of the phase on time can be used for excluding the refractive attenuation from an amplitude function of the SPMM and FIO spectra. However, as described in detail [77], there exist significant difficulties to measure the atmospheric absorption of radio waves by FSI and other radio holographic methods based on a Fourier operator. An important connection between the eikonal acceleration, Doppler shift, phase, and intensity variations of RO signals has been revealed by theoretical considerations and experimental analysis of the ionospheric parts of the RO radio holograms [55,82,83,136-138]. They introduced an eikonal acceleration technique which gives a simple way to convert the phase data into the refractive attenuation. This is useful for estimating the total absorption of radio waves in atmospheric communication links. In this section the eikonal acceleration technique is validated by use of the CHAMP GPS data, and a method for identification of layered structures in the atmosphere and ionosphere is presented. New relationship between the bending angle and refractive attenuation is presented. A possibility to measure the total absorption of radio waves at GPS frequency by use of proposed eikonal acceleration technique is considered.

4.6.1. Identification of layered structures and estimation of the refractive attenuation from the phase data

The geometry of the GPS radio occultation experiment is shown in Fig. 3.4.1. The radio waves emitted by a GPS satellite (point G) are propagating to a LEO satellite (point L) along the ray trajectory GTL, where T is the ray perigee. The projection of point T on the surface of the Earth determines the geographical coordinates of the RO session. Connections between the eikonal acceleration Φ and refractive attenuation X of radio waves have been detected in [55,82,83] (Eqs. (3.4.15)). Eqs. (3.4.15) connect the refractive attenuation $X(t)$, derivative of the Doppler frequency F_d on time and the eikonal acceleration $a_\Phi = \frac{dF_d}{dt} = \frac{d^2\Phi}{dt^2}$ via a relationship similar to classical dynamics equation. Eqs. (3.4.15) are based on an exact formula for the refractive attenuation found from an energy conservation principle [53], and are valid in general case of strong refractive effect. Parameters m and dp_s / dt may be evaluated from the orbital data. Therefore, Eqs. (3.4.15) give a possibility to convert the eikonal acceleration a_Φ and/or Doppler shift F_d to the refractive attenuation X_p. Another important equation is valid for derivative $\frac{d\xi}{dt}$ and followed from a 2-D formula for the refractive attenuation X_a

$$X_a = \frac{1}{1 - q\frac{d\xi}{dp}}, \quad \frac{d\xi}{dp} = \frac{d\xi}{X_a dt}\left(\frac{dp_s}{dt}\right)^{-1}, \quad \frac{d\xi}{dt} = (X_a - 1)\frac{dp_s}{qdt}. \tag{4.6.1}$$

A ratio X_a of the intensities of radio signal propagating through the atmosphere, $I_a(t)$, and free space I_s, (Eq. (3.4.18) is equal to product of the refractive attenuation X_p, absorption losses Y, and may include technical instabilities. Comparison of values of X_a and X_p (curves 1 and 2) in the altitude interval 5–30 km are shown for CHAMP RO event 0033 in Fig. 4.1.2. The standard dependence of the combined atmospheric refractive attenuation and absorption is indicated in Fig. 4.1.2 by curves 3. To obtain the refractive attenuation X_p the eikonal acceleration a_Φ has been estimated as a second derivative of the phase path excess with respect to time t over a fixed time interval Δt, which was equal to 0.42 s. Values X_a and X_p change in similar manner in the stratosphere (Fig. 4.1.2). This identifies a common origin of the phase and intensity variations of RO signal as a contribution of layered structures in the atmosphere and ionosphere. As a further identification step a possibility to locate layers in the atmosphere and/or in the ionosphere may be examined. In the absence of noise and absorption parameter m may be evaluated from (4.6.1), and then the distance d_{2s} can be estimated from the relationship (3.4.21), (3.4.22). Then it is possible to determine the displacement $d = d_{2s} - (R_2^2 - p_s^2)^{1/2}$ of a layer relative to the point T (Fig. 3.4.1). Results of determination of the parameter m, displacement d, and corrected layer's height h are given in the Table 4.6.1 as functions of the altitude H of ray perigee T. Data indicated in the Table 4.6.1 correspond to CHAMP GPS RO event № 0033. Values X_a, X_p calculated from the amplitude and phase data, and estimated values of parameter m are indicated in the second, third, and sixth columns in the Table 4.6.1 in the altitude interval of 9.26–9.36 km. The displacement d and an estimated value of the layer's altitude h are indicated in the forth and fifth columns of the Table 4.6.1, respectively. The displacement d is changing in the 6 to 40 km interval and corresponding corrections to the altitude H are about 0.1–0.2 km in average. These results confirm location of the atmospheric layers near the ray perigee T.

H, km	X_p	X_a	d, km	h, km	δ °	m, s²/m
9.36	0.32461	0.29659	–5.35	9.36	– 0.046	0.43083
9.35	0.32266	0.29152	4.98	9.35	0.044	0.43308
9.33	0.32062	0.28662	14.4	9.35	0.128	0.43515
9.32	0.31833	0.28197	22.3	9.36	0.200	0.43687
9.31	0.31594	0.27761	29.0	9.37	0.258	0.43831
9.30	0.31341	0.27354	34.2	9.39	0.306	0.43945
9.28	0.31069	0.26984	37.7	9.40	0.338	0.44021
9.27	0.30773	0.26645	39.6	9.40	0.354	0.44060
9.26	0.30407	0.26334	38.5	9.38	0.344	0.44035

Table 4.6.1 Displacement d in the neutral atmosphere

4.6.2. Determination of total absorption

Note that the refractive attenuation $X(t)$ in Eqs. (3.4.15) depends only on the atmospheric refraction effect and ignores the atmospheric absorption. Usually the parameters m and dp_s / dt are known from orbital data because location of the spherical symmetry center O and its projection on the line of sight- point D - are known, and the distance $GD \equiv d_{1s}$ and $DL \equiv d_{2s}$ can be easily estimated from trajectory data. Therefore, Eqs. (3.4.15) open a possibility to convert the eikonal acceleration a_Φ and/or Doppler shift F_d to the refractive attenuation. The refractive attenuation X_a is determined via Eq. (3.4.18) from the amplitude data as a ratio of intensities of radio signal propagating through the atmosphere, $I_a(t)$, and free space, I_s. The experimental value X_a is the result of the contributions from refraction and absorption effects. This provides a possibility to determine the total atmospheric absorption $Y(t)$ as the ratio where $X_p(t)$ is the refractive attenuation of radio waves recalculated from the phase data. The estimated value $X_p(t)$ contains only the contribution from refraction effects. Therefore Eqs. (4.6.2) and (4.6.3) allow one to exclude the refractive contribution and to estimate the contribution from the total absorption.

$$Y(t) = X_a(t) / X_p(t), \tag{4.6.2}$$

with

$$X_p = 1 - ma_\Phi = 1 - m\frac{dF_d}{dt}, \tag{4.6.3}$$

Analysis of the CHAMP RO data indicates that the relationship (3.4.15) is correct. A typical example from RO session 0096 conducted on November 20, 2003, is shown in Fig. 4.6.1. Vertical profiles of the atmospheric phase path excesses measured at frequencies L1 and L2 are indicated on a logarithmic scale by curves 1 and 2 (left panel). For convenience an artificial bias of 1 m was introduced into the phase path excesses. The phase path excesses L1 and L2 change from 2-5 m at 110 km to 1000 m at 1 km height. Curve 3 describes the altitude dependence of contribution of the neutral atmosphere, isolated from the ionospheric impact by use of linear combination of the phase path excesses L1 and L2. This neutral atmosphere phase path excess changes over a broad dynamic range from 1 mm at 75 km to 1 km near the earth's surface. In contrast, the refractive attenuation X_a changes in a steep dynamic range from 0.05 to 1.1 over the height range 4 km to110 km (curve 1, right). Despite this difference, the relationships (3.4.15) bring to light an intrinsic connection between the phase path excess and intensity of RO signal.

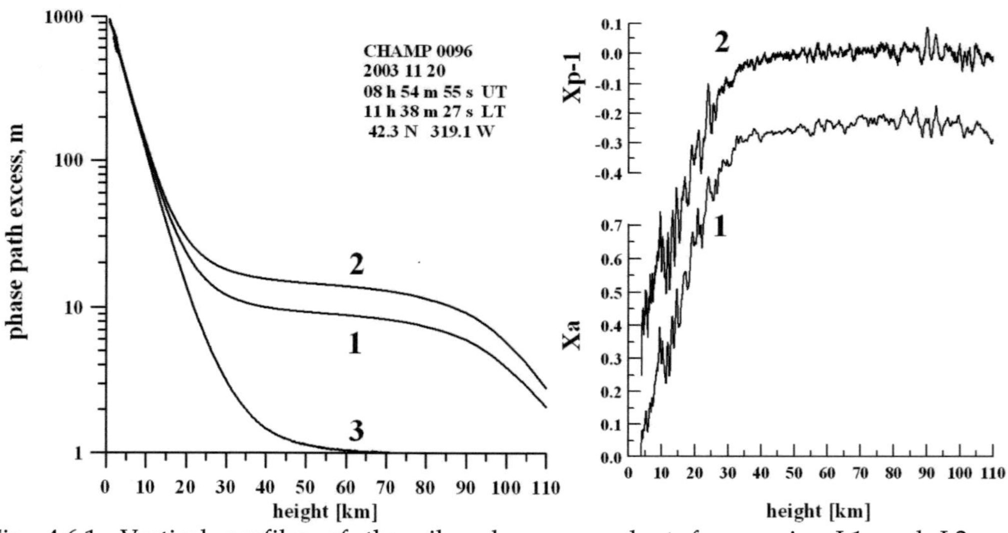

Fig. 4.6.1. Vertical profiles of the eikonals measured at frequencies L1 and L2 and recalculated by use of ionospheric correction (curves 1,2, and 3, respectively, left panel). Vertical profiles of the refractive attenuations X_a and X_p computed from the amplitude and phase data (curves 1 and 2, respectively, right panel).

This may be seen from comparison of the refractive attenuations X_a and X_p computed from the amplitude and phase data (right, curves 1 and 2). In order to convert the phase data according to equation (3.4.15), the eikonal acceleration a_Φ has been estimated numerically as the second derivative of the phase path excess with respect to time t by use of a fixed time interval value of $\Delta t = 0.42$ s for differentiation. As a result of differentiation the high frequency noise level increases. This effect is seen in Fig. 4.6.1 (right). The relationships (3.4.15) widens the applicability domain of RO method. In particular RO method can be applied for estimating the total absorption of radio wave in the atmosphere. Vertical profiles of the total absorption of radio wave in the atmosphere are shown in Fig. 4.6.2 for five CHAMP RO measurement sessions conducted on November 20, 2003. The sessions 0131 and 0175 correspond to polar and mid latitude areas in the Southern hemisphere (panels a and c); session 0187 refers to an equatorial region (panel b). The session 0096 was carried out in area of mid latitude in the Northern hemisphere. Therefore these measurements represent conditions of radio wave propagation in all typical regions of the Earth. The refractive attenuations X_a and X_p recalculated from the amplitude and phase data using Eqs. (4.6.2) and (4.6.3) are shown on the left panels (curves 1 and 2). The smooth curves 3 indicate dependence of the refractive attenuation corresponding to the exponential altitude profile of the refractivity in the atmosphere with accounting for the total path absorption effect calculated by using the theoretical and experimental results [128] for decimeter-range radio waves. Excellent correspondence is seen between the refractive attenuations X_a and X_p changing from 0 dB at 40 km to –10 dB and –15 dB at 5 km. There exists also a good correlation between the high-frequency part of variations in X_a and X_p. However in Fig. 4.6.2 (left, c) good correspondence between the high-frequency part of variations in X_a and X_p exists only over an interval 5–20 km. Above 20 km the high-frequency variations of the refractive attenuation X_a are significantly greater than those of X_p. This may be connected with a possible effect of transmitter-receiver instability and/or with ionospheric variability . Most probable cause is influence of the polar ionospheric plasma structures originated due to an impact of the intense geomagnetic storm occurred on November 20, 2003. The magnitude of the scintillation index S_4 equaled 13.7% in the polar session 0131. The magnitude of the scintillation index S_4 was 5.5% (middle latitude, session 0096, left panels, a), and 4.8% in the equatorial area (session 0187, left panel, b). These values for the scintillation index S_4 are higher than those for quiet ionospheric conditions when S_4 changes mainly between 1% and 2% due to influence of the receiver noise. Therefore the geomagnetic storm introduced significant contribution in the amplitude and phase variations of RO signal which is clearly seen in the polar regions of the Earth.

Apparent connection between the refractive attenuations restored from the amplitude and phase variations is useful for estimating the altitude dependence of the total absorption in the atmosphere. This dependence is demonstrated by the rough curves 1 in Fig. 4.6.2 (right panels). The smooth curves 2 correspond to the total absorption due to atmospheric oxygen calculated according to techniques described in [29, 128]. Theoretically, influence of the atmospheric oxygen is noticeable below 10–15 km. Experimental data qualitatively support this theoretical suggestion. Below altitude 8 km experimental data indicate additional attenuation as compared with theoretical dependence. Also influence of additional absorption in clouds and water vapor may be important below 8 km. For quantitative estimation of small absorption values in the decimeter frequency band it is necessary to apply averaging for analysis of the experimental data.

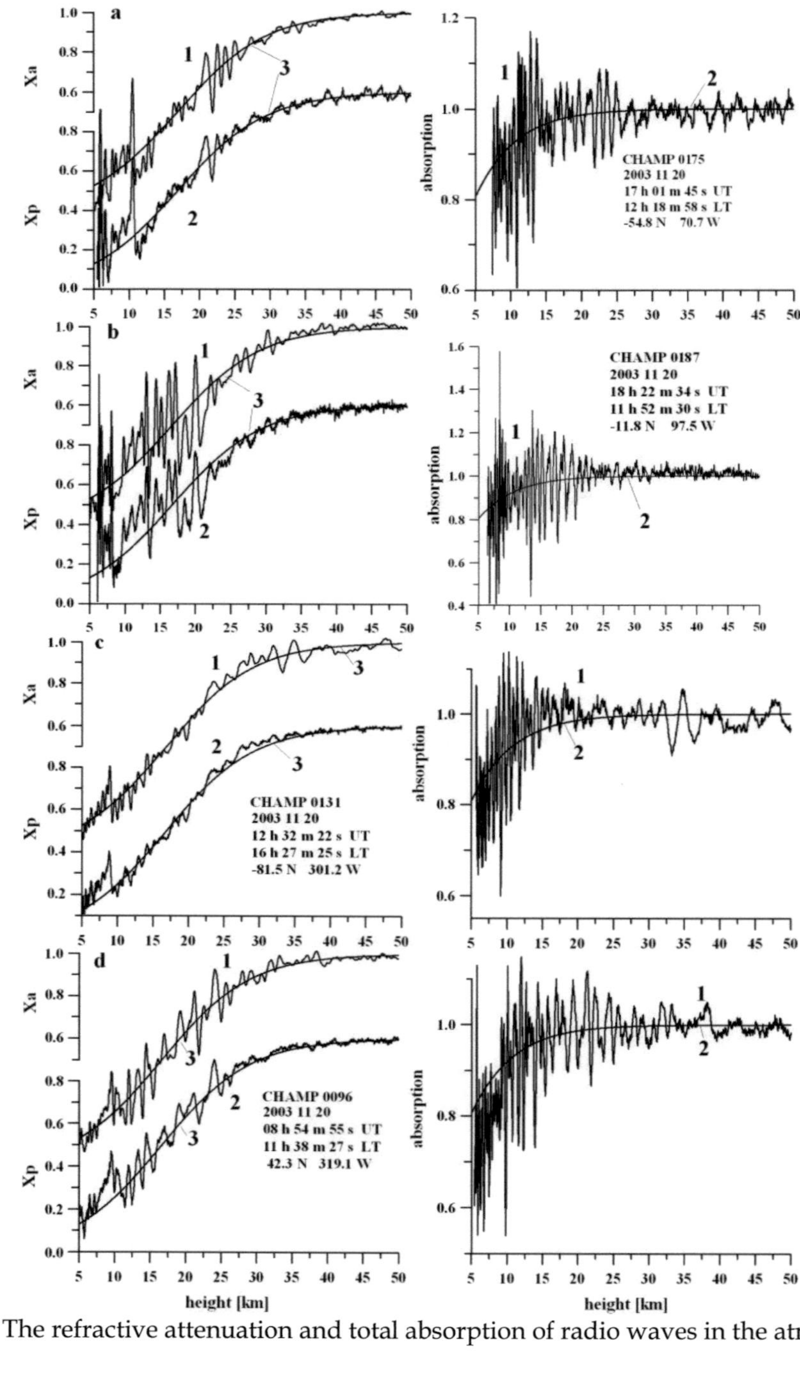

Fig. 4.6.2. The refractive attenuation and total absorption of radio waves in the atmosphere.

To obtain smoothed dependence $Y(h)$ the vertical profiles of $X_a(h)$ and $X_p(h)$ have been approximated by polynomials using a least square method. In Fig. 4.6.3 vertical profiles of the refractive attenuation approximated $X_a(h)$, $X_p(h)$, and absorption $Y(h)$ are shown by curves 1-5 and 6-10, respectively, for five CHAMP RO measurement sessions conducted in November 20, 2003. For convenience an artificial bias of 0.2 was introduced in consecutive order into the curves 2-5, 6-9, and B. Dashed curves A and B correspond to vertical profiles of $X_a(h)$ and $Y(h)$ evaluated from standard atmospheric model. The total absorption due to atmospheric oxygen $Y(h)$ has been calculated by use of technique described in [128]. At the altitudes between 12 and 30 km vertical profiles $X_a(h)$ and $X_p(h)$ are nearly coinciding and have good correspondence with the standard profile (curve B). Below 12 km altitude they begin split (e.g., curves a and p) at different heights. The splitting exists practically in all RO events. Difference in the splitting altitude may be connected with an amplitude instability. The splitting obviously indicates an influence of the atmospheric total absorption, which in average is near to values (curves 6-10 and curve B), described in [29–31, 128]. Absorption is changing between 0–30% in the altitude interval 12 - 5 km (Fig. 4.6.3, curves 6-10). Value of the amplitude instability error in measured absorption is estimated as $\pm$ 20% from maximal value. The described results indicate a possibility of the RO total absorption measurements at GPS frequencies.

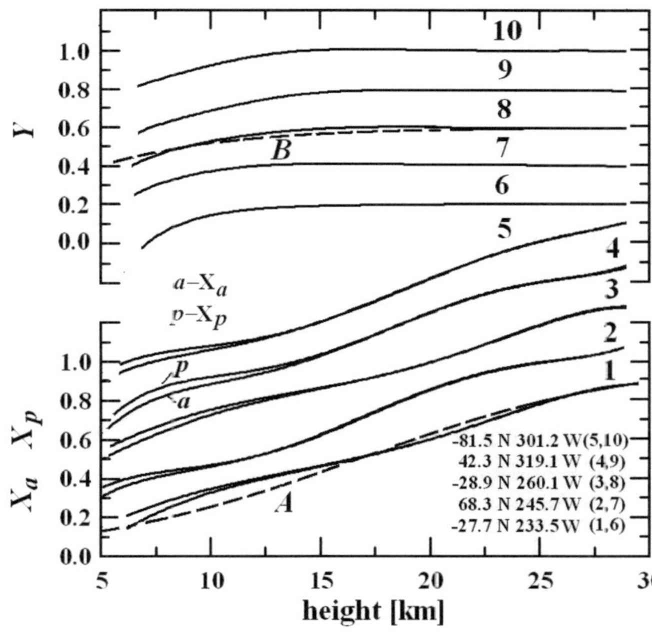

Fig. 4.6.3. Vertical profiles of $X_a(h)$, $X_p(h)$, and $Y(h)$ (CHAMP RO data). Inserts indicate the geographical coordinates of investigated regions.

In this section two connections between the eikonal acceleration, derivative of the bending angle with respect to time and the refractive attenuation variations is validated by means of the theoretical consideration and experimental analysis of data registered during CHAMP mission. These connections give a possibility to convert the eikonal acceleration (or time derivative of Doppler shift) into the refractive attenuation, and allows one to measure the total absorption of radio waves in the trans-atmospheric communication links. The eikonal acceleration technique may be used to identify and locate layered structures in the earth's and planetary atmospheres and ionospheres including the case of significant horizontal gradients. In some instances it is possible to establish the location, height, and inclination of E_s structures in the ionosphere from a single RO vertical profile. It follows from our analysis that the refractive attenuations restored from the amplitude and phase variations are useful for estimating the total absorption, recognizing layered structures in the atmosphere and revealing the systematic errors in the amplitude data.

4.7. Investigation of radio wave fluctuations and small-scale atmospheric irregularities

The fluctuations of the refractive index $\delta n(r,t)$ are always present in the atmosphere. These fluctuations are affecting on radio waves propagation. Movement of radio ray through realization of a random function $\delta n(r,t)$ introduces oscillations of the amplitude and phase. Solution of problem of wave propagation through inhomogeneous medium contemplates an ascertainment of connections between the statistical characteristics of the field of random function $\delta n(r,t)$ and statistical parameters of wave. In the beginning let us to introduce the initial statements, and then to describe according to the monograph [139] the basic results of the theory of wave propagation in a random inhomogeneous medium.

Let us remind the main definitions and relationships between the statistical characteristics of the fluctuations $\delta n(r,t)$: spatial correlation function $\mathrm{B_n}(\rho)$, structure function $\mathrm{D_n}(\rho)$ and spatial spectrum $\Phi_\mathrm{n}(\mathrm{K})$. Let us assume, that the fluctuation's field δn is isotropic and stationary, i.e. $\mathrm{B_n}(\rho)$, $\mathrm{D_n}(\rho)$ and $\Phi_\mathrm{n}(\mathrm{K})$ do not depend on the direction and time. Spatial correlation function of the fluctuations of refractive index may be determined by the next way

$$\mathrm{B_n}(\rho) = \left\langle \delta\, n(\mathbf{r}_1,t)\cdot \delta\, n(\mathbf{r}_2,t)\right\rangle , \tag{4.7.1}$$

where r_1 and r_2 are the coordinates of two points C_1 and C_2, ρ is the distance between C_1 and C_2, and the brackets $\langle\ \rangle$ signify averaging on different realizations δn. In Eq. (4.7.1) the fluctuations δn have in average zero value, since a smooth trend of refractive index may be excluded. From (4.7.1) it follows, that $\mathrm{B_n}(\rho = 0)$ is equal to the dispersion of fluctuations of refractive index $\sigma_\mathrm{n}^{\ 2}$, and $\mathrm{B_n}(\rho \to \infty) = 0$. Correlation function $\mathrm{B_n}(\rho)$ characterizes the statistical connection of fluctuations δn in the points C_1 and C_2. If irregularities are caused by developed turbulence, then

$$\mathrm{B_n}(\rho) = \sigma_n^2 \left[1 - \left(\frac{\rho}{\Lambda_\mathrm{m}}\right)^{2/3} \right] \quad \text{when } \rho \le \Lambda_\mathrm{m} , \tag{4.7.2}$$

where Λ_m is the external scale of irregularities. By applying the correlation function $B_n(\rho)$, one may introduce the effective scale of irregularities

$$\Lambda_e = \sigma_n^{-2} \int_0^{\infty} B_n(\rho)\, d\rho \ . \qquad (4.7.3)$$

If B_n corresponds to Eq. (4.7.2), then effective scale is equal to $\Lambda_e = \frac{2}{5}\Lambda_m$. For statistical description of the field δn one may apply also the structure function D_n, which may be determined as $D_n(\rho) = \left\langle [\delta\, n\,(\mathbf{r}_1,t) - \delta\, n\,(\mathbf{r}_2,t)]^2 \right\rangle$.

If fluctuations of the refractive index may be considered as isotropic and stationary, then usage of function $B_n(\rho)$ or $D_n(\rho)$ is limited only by convenience of calculations, they are connected in this case by a linear relationship. The structure function in many cases may be satisfactorily approximated by a power function

$$D_n(\rho) = C_n^{\ 2} \rho^q \ . \qquad (4.7.4)$$

In the case of turbulent gas q=2/3. The structure function (4.7.4) corresponds to the correlation function (4.7.2) when q=2/3. Dependences (4.7.2) or (4.7.4) are valid, if the distance ρ is limited by inequality $\Lambda_m >> \rho >> \Lambda_0$, where Λ_0 is a small internal scale. Parameter C_n is the structure characteristic of the refractive index fluctuations, it is connected with dispersion $\sigma_n^{\ 2}$ and external scale Λ_m; if q=2/3 then

$$C_n^2 = 2\ \sigma_n^2\ \Lambda_m^{-7/3} \ . \qquad (4.7.5)$$

Spectral approach is effective for description of the field fluctuations δn. If the arbitrary direction OX is selected in a medium then because isotropy and homogeneity of the field δn the random function of the variables x and t may be obtained, and one will have the corresponding correlation function $B_n(\rho)$, where $\rho = x_1 - x_2$. The correlation function B_n corresponds to 1-dimensional spectrum of the refractive index fluctuations $V_n(K_x)$. The function $B_n(\rho)$ and 1-dimensional spectrum $V_n(K)$ are connected by the Fourier-transformation

$$B_n(\rho) = \int_{-\infty}^{\infty} \cos\,(K_x\rho)\ V_n(K_x)\ dK_x \ , \qquad (4.7.6)$$

$$V_n(K_x) = \frac{1}{2\pi} \int_{-\infty}^{\infty} \cos\,(K_x\rho)\ B_n(\rho)\ d\rho \ . \qquad (4.7.7)$$

Here $K = 2\pi\Lambda_x^{-1}$ is the wave number of irregularities in the medium, and Λ_x is the corresponding horizontal scale. The spectrum $V_n(K_x)$ describes distribution of the intensity of fluctuations δn as function of the spatial scale Λ_x. The 3-dimensional spectrum $\Phi_n(K)$ of the fluctuation's field δn is determined by means of generalization of (4.7.7) upon three space coordinates

$$\hat{O}_n(K) = \frac{1}{(2\pi)^3} \int\int_{-\infty}^{\infty} \int \cos(\mathbf{K}\rho)\, B_n(\rho)\, dx\, dy\, dz, \tag{4.7.8}$$

where $K\rho = K_x\rho_x + K_y\rho_y + K_z\rho_z$, $K^2 = K_x{}^2 + K_y{}^2 + K_z{}^2$, $K = 2\pi\Lambda^{-1}$ is the spatial wave number, Λ is the spatial scale. If the field δn is homogeneous and isotropic, then one may conduct in (4.7.8) the integration in the spherical system on the angular coordinates and arrive to the more simple connection between B_n and Φ_n

$$\hat{O}_n(K) = \frac{1}{2\pi^2 K} \int_0^{\infty} \rho\, B_n(\rho) \sin(K\rho)\, d\rho. \tag{4.7.9}$$

The next connection exists between the 1-dimensional and 3-dimensional spectrums

$$\Phi_n(K) = -\frac{1}{2\pi\, K} \frac{d\, V_n(K)}{dK}, \tag{4.7.10}$$

which allows finding the 3-dimensional spectrum Φ_n, if the 1-dimensional spectrum of the fluctuations of the refractive index V_n is known. By letting in (4.7.6) $\rho = 0$ one obtains the relationship connecting the dispersion of the refractive index fluctuations $\sigma_n{}^2$ with the 1-dimensional spectrum of irregularities v_n

$$\sigma_n^2 = B_n(0) = 2 \int_0^{\infty} V_n(K)\, dK. \tag{4.7.11}$$

A connection of the dispersion $\sigma_n{}^2$ with the 3-dimensional spectrum Φ_n is determined by the next formula

$$\sigma_n^2 = B(0) = 4\pi \int_0^{\infty} K^2\, \Phi_n(K)\, dK. \tag{4.7.12}$$

The lower limit from the formal point of view is equal to zero in (4.7.11) and (4.7.12); actually the spectrum $\Phi_n(K)$ is known only for $K \geq K_m$. From (4.7.7) it follows, that if the correlation function is presented by the relationship (4.7.2) the spectrum is equal to

$$V_n(K) = \frac{\Gamma(5/3)}{2\pi} \sin\left(\frac{\pi}{3}\right) C_n^2 K^{-5/3} , \tag{4.7.13}$$

where Γ is the gamma-function. Since $K = 2\pi\Lambda^{-1}$, then from Eq. (4.7.13) it follows, that the spectral density v_n has the maximal values in the range of large scales of the irregularities, and for high-frequency part of spectrum, when $\Lambda \to 0$, one has $V_n \to 0$. Theory of random function contends, that if the structure function D_n may be presented by a power function (4.7.4), then the spectrum Φ_n also will be a polynomial: $\Phi_n \sim K^{-(q+3)}$. In the case of developed turbulence, when presentation of the structure function by Eq. (4.7.4) where *q*=2/3 is valid, one obtains the 3-dimensional Kolmogorov' spectrum

$$\Phi_n(K) = 0.033\ C_n^2 \left(K^2 + K_m^2\right)^{-11/6} , \tag{4.7.14}$$

where $K_m = 2\pi\Lambda_m^{-1}$. One may apply general expression (4.7.12) to find the dispersion of fluctuations σ_n^2 and Eq. (4.7.14) to define the spectrum of irregularities and determine connection between C_n and σ_n; in the case of the Kolmogorov' spectrum one may obtain the relationship (4.7.5).

Spatial spectrum of the refractive index fluctuations $\Phi_0(K)$ depends on the physical processes, begetting or cancelling the fluctuations δn. Experimental investigations indicated that, despite the difference of these processes, it is possible to get an approximated analytical presentation of the spectrum Φ_n, which is valid for different media. It is turned out, that for a large interval of changes of the spatial wave number this spectrum may be approximated by the functions

$$\Phi_n = 0.033\ C_n^2 \left(K^2 + K_m^2\right)^{-q/2} \exp\left(-\frac{K^2}{K_0^2}\right) , \tag{4.7.15}$$

or

$$\Phi_n \approx 0.033\ C_n^2 \left(K^2 + K_m^2\right)^{-q/2} , \tag{4.7.16}$$

where q is the index of the 3-dimensional spatial spectrum of fluctuations δn, $K_m = 2\pi\Lambda_m^{-1}$ and $K_0 = 2\pi\Lambda_0^{-1}$. Approximated relationship (4.7.16) corresponds to the large part of spectrum, where $K < K_0$. The conditional external scale Λ_m depends on the processes, begetting the statistical irregularities of a medium, and the scale Λ_0 is connected with the phenomenon of cancelling the irregularities and leading to the sharp decrease of the spectral density Φ_n when $K > K_0$. The scales Λ_m and Λ_0 distinguish strongly for different media, however the index of spatial spectrum q changes in a short range. Investigation indicated, that the spectral index q for the troposphere is near to the theoretical Kolmogorov' value $q = 11/3$, corresponding to a medium with developed turbulence [139]. For the stratosphere the spectral index may differ from $q = 11/3$ because of influence of the wave phenomena on formation of the spectrum $\Phi_n(K)$. In addition irregularities may be anisotropic in the stratosphere: the horizontal scale is usually many times greater of vertical sizes of irregularities. Since the spectral density of fluctuations when $\Lambda < \Lambda_0$ is small, then this part of spectrum does not get any notable influence on fluctuation's characteristics of radio waves; therefore one may conclude that approximated spectrum (4.7.16) is valid also for small scales. When $\Lambda > \Lambda_m$ the spectrum Φ_n becomes uncertain, and irregularities are always strongly anisotropic, therefore effect of influence on radio wave of the large-scale irregularities, with the sizes greater the external scale Λ_m is difficult for analysis.

If radio ray is moving in a medium, then in the different instants of time t the different realizations of $\delta n(r,t)$ will intersect the radio ray and, in general case, one should consider the complex temporal-spatial statistical relationships. One may avoid these difficulties, if suppose, that a statistically inhomogeneous medium during some time interval Δt remains invariable or «frozen». This «frozen» hypothesis is valid, if the irregularities having the scale Λ are intersecting by radio ray during the time $\Delta t = \Lambda v^{-1}$, which should be significantly smaller the time of essential changes of these irregularities. In problem under consideration this condition takes place always. In the simplest case the statistically inhomogeneous medium is intersecting by the velocity vector v perpendicularly to the direction of radio wave propagation LG. Under assumption of «frozen» irregularities the fluctuations δn in a point with coordinate y in the time instant t coincide with fluctuations in the point $(y - vt)$ in the time instant $t = 0$ (Fig. 4.7.1). Hence, the condition of «frozen» irregularities may be presented by equation $\delta n(r,t) = \delta n(r - vt, 0)$. This relationship allows avoiding the consideration of the complex temporal-spatial statistical laws that are guiding variations of the field δn and allows usage of the medium characteristics B_n and Φ_n.

Let us consider a connection of fluctuations δn with random changes of the atmospheric temperature δT_a. From (2.1.3) the next relationships may be obtained

$$\delta n = -\chi_1 P_a T^{-2} \delta T \quad \text{or} \quad \sigma_n^2 = \chi_1^2 P_a^2 T^{-4} \sigma_t^2 , \qquad (4.7.17)$$

where σ_t^2 is the dispersion of the temperature fluctuations, $\chi_1 = 77.6$ if the pressure P_a is expressed in millibars, and temperature T is measured in Kelvin. Influence of the humidity fluctuations does not be accounted for in Eq. (4.7.17). This influence may be significant in the boundary layer of the atmosphere.

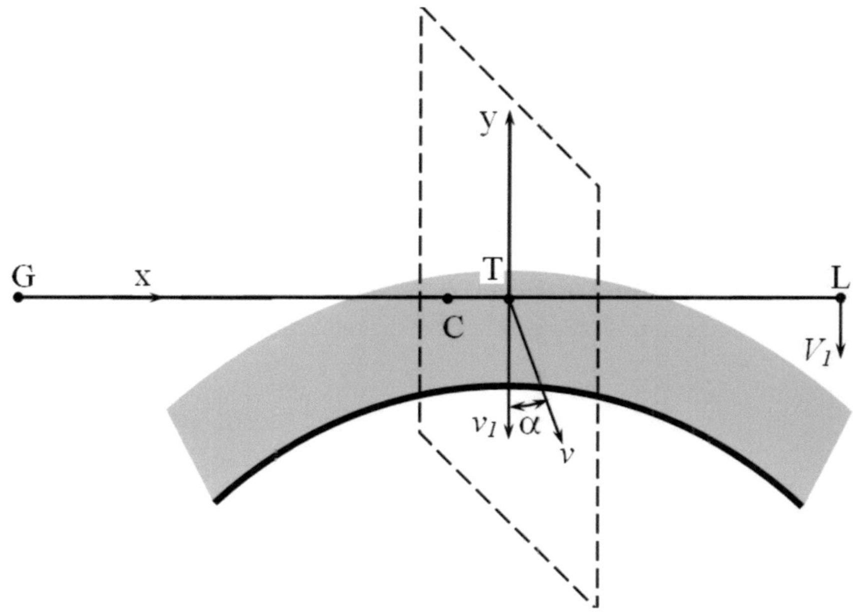

Fig. 4.7.1. Geometry of problem of radio wave propagation through irregular medium.

Let us consider results of theory of wave propagation in the statistically inhomogeneous media. Let us suppose for simplicity that the width of an inhomogeneous medium along the wave propagation direction is equal to Δx. Then the dispersion of the phase fluctuations σ_ϕ^2 according to [139] is determined by relationship

$$\sigma_\phi^{\ 2} = 8\pi^2\lambda^{-2}\sigma_n^{\ 2}\Lambda_e\Delta x\,, \tag{4.7.18}$$

where Λ_e is the effective scale of irregularities determined by Eq. (4.7.3). In the case when Eqs. (4.7.2) and (4.7.14) are valid $\Lambda_e = 2/5\Lambda_m$. If the structure characteristic C_n is used then according to (4.7.5) the next equivalent expressions should be introduced instead of σ_n

$$\sigma_\phi^{\ 2} = \frac{16\pi^2}{5\lambda^2}\,\sigma_n^{\ 2}\Lambda_m\,\Delta x\,, \tag{4.7.19}$$

$$\sigma_\phi^{\ 2} = \frac{8\pi^2}{5\lambda^2}\,\tilde{N}_n^{\ 2}\Lambda_m^{\ 5/3}\,\Delta x\,. \tag{4.7.20}$$

As follows from theory the frequency spectrum of the phase fluctuations is described by the power function

$$\hat{O}_\phi \sim F^{\ -m}\,, \tag{4.7.21}$$

$$m = q - 1\,, \tag{4.7.22}$$

where F is the frequency of fluctuations which is connected with the velocity v of medium intersection by the radio ray owing to the approximated relationship $F = \Lambda^{-1} v$, and q is the index of the spatial spectrum of the refractivity fluctuations. From Eqs. (4.7.19) or (4.7.20) it follows, that σ_ϕ is inversely proportional to the wavelength and depends, mainly, on the large-scale component of spectrum of the refractivity fluctuations. It is necessary to note that these relationships, describing the phase fluctuations, are valid, if the scale of irregularities Λ is smaller the external scale Λ_{m}.

According to [139] the dispersion of the amplitude fluctuations σ_a^2 corresponds to the developed turbulence and may be expressed by the relationship when $q = 11/3$

$$\sigma_{\grave{a}}^{\,2} = 0.56\ \hat{e}^{7/8}\ \mathrm{L}_l^{-5/6} \int_0^l \mathrm{C}_{\mathrm{n}}^{\,2}(\mathrm{x})\ \left[\mathrm{x}\ (\mathrm{L}_l - \mathrm{x})\right]^{5/6}\ \mathrm{dx}\ , \qquad (4.7.23)$$

here $\kappa = 2\pi / \lambda$, L_l is the length of the distance LT (Fig. 2.1.1). If irregularities are located in the layer having the width $\Delta\mathrm{x}$, then it follows from Eq. (4.7.23)

$$\sigma_a^{\,2} = 0.56 \hat{e}^{7/6}\ \tilde{\mathrm{N}}_{\mathrm{n}}^{\,2}\ \Delta\mathrm{x} \left(\frac{\mathrm{L}_l \mathrm{L}_{\mathrm{g}}}{\mathrm{L}_l + L_{\mathrm{g}}} \right)^{5/6}, \qquad (4.7.24)$$

where $\mathrm{L}_l = \mathrm{LT}$, $\mathrm{L}_g = \mathrm{GT}$ are the distances from receiver and transmitter from the turbulent layer, respectively (Fig. 4.7.1). It follows from Eq. (4.7.24), that the dispersion of the amplitude fluctuations depends on the distances L_l and L_g and is increasing with the wavelength decreasing according to the law $\sigma_a^2 \sim \lambda^{-7/6}$. It is indicated in [139], that the frequency spectrum of the amplitude fluctuations $\Phi_a(F)$ has two characteristics regions: when the frequency of fluctuations $F < F_0$ the spectral density is independent on frequency, and when $F > F_0$ the spectral density is diminishing with increasing of the frequency of fluctuations according to the power law

$$\Phi_a \sim F^{-m}\ , \qquad (4.7.25)$$

where m is connected with the index of the spatial spectrum of the medium irregularities q by Eq. (4.7.22). The characteristic frequency F_0 depends on the size of the first Fresnel' zone, i.e. on L_l, L_g, λ and velocity v

$$\mathrm{F}_0 = \mathrm{v} \left[\frac{\mathrm{L}_l + \mathrm{L}_{\mathrm{g}}}{2\pi\ \lambda\ \mathrm{L}_l\ \mathrm{L}_g} \right]^{1/2}. \qquad (4.7.26)$$

Theory of the amplitude fluctuations is valid for the relatively small intensity of fluctuations $\sigma_a < 0.6$. It follows from considered relationships describing the dispersions of the amplitude and phase fluctuations, that σ_a or σ_φ may be employed as the characteristics of intensity of small-scale atmospheric irregularities, and the spectra Φ_a and Φ_a give an information concerning the features of the spectrum of the refractivity fluctuations Φ_{n}.

Let us consider below the scheme of radio occultation of the statistically inhomogeneous atmosphere in the communication link satellite-to-satellite indicated as the straight line LG in Fig. 4.7.1. The straight line LG has at point T the minimal height H relative to the earth surface. Irregularities, located on the radio ray LG in an arbitrary point C, have the different heights h and therefore it is necessary to account for dependence $\sigma_{\mathrm{n}}(h)$ or C_n on the height. Having analogy with papers [140, 141], where the different dependences of $\sigma_{\mathrm{n}}(h)$ were considered, let us suppose, that in the earth's atmosphere the fluctuations σ_{n}^2 are diminishing with height according to an exponential law

$$\sigma_n^2 = \sigma_0^2 \ exp(-\ 2\beta h), \tag{4.7.27}$$

here σ_0^2 is the dispersion of refractivity fluctuations near the earth' s surface. Since the irregularities, located on the radio ray LG, are independent, then σ_φ and σ_a may be found by means of integration of the right parts of Eqs. (4.7.19), (4.7.20) and (4.7.24) along the radio ray. Let us introduce the coordinate axis OX with center in the point T as is shown in Fig. 4.7.1 and to conduct the integration with accounting for dependences (4.7.27) and relationships (4.7.5)

$$\sigma_\phi^2 = \frac{16\pi^2}{5\lambda^2}\, \Lambda_{\mathrm{m}} \left(\frac{\pi\ a}{\beta}\right)^{1/2} \sigma_0^2 \exp\left[-\ 2\ \beta H\right], \tag{4.7.28}$$

$$\sigma_a^2 = \hat{e}^{7/6}\ \Lambda_{\mathrm{m}}^{-2/3} \left(\frac{\pi\ a}{\beta}\right)^{1/2} \left(\frac{L_l L_g}{L_l + L_g}\right)^{5/6} \sigma_0^2 \exp\left[-2\ \beta H\right], \tag{4.7.29}$$

where H is the minimal altitude of radio ray found with accounting for the refraction of radio waves, and $(\pi a \beta^{-1})^{1/2}$ is the width of a conditional «layer» of a statistically inhomogeneous medium, which causes the same fluctuations as a spherical symmetric inhomogeneous medium. The altitudes in the atmosphere are many times smaller the Earth radius a, and, therefore, an approximated equation can be used

$$\mathrm{x} \approx \left[2a(a-H)\right]^{1/2}.$$

Let us go to description of results of experimental investigations of radio waves fluctuations and atmospheric irregularities. These investigations were accomplished with usage of the communication links orbital station MIR– geostationary satellites and GPS– Microlab[142–145]. The amplitude fluctuations of centimeter radio waves will be analyzed at first following to papers [29, 142–145]. Experiments were conducted by the next way: radio waves with wavelength $\lambda = 2$ cm were emitted by the orbital MIR station and passed through the atmosphere. Radio waves were received onboard of a geostationary satellite and then were retranslated to the ground based station, where the registration of signal was provided . Measurements series contain 6–8 sessions and were carried out during different seasons of year, daytime and above different regions with usage of two geostationary satellites: the first satellite was located in Western hemisphere, and the second one – in Eastern hemisphere. The orbital inclination of MIR station equal to 52° allows providing RO sounding of the atmosphere in the Southern and Northern hemispheres in the latitude interval 0–52°. Investigations were accomplished above the Indian Ocean, above the Northern part of the Pacific Ocean and also above Kazakhstan. When analyzing the amplitude fluctuations it is necessary to know the distances L_l and L_g of the investigated region of the atmosphere from transmitter and receiver, respectively. The distances L_l and L_g have the next values: L_l=2,100–2,500 km and L_g=41,700 km. Knowledge of the velocity of the irregularities transfer across the communication links is required for analysis of spectra of the amplitude and phase fluctuations. Since the geostationary satellites are motionless, the irregularities transfer occurs because of movement of the MIR station.

The ray line GL is always located in the plane passing through transmitter, receiver and the earth' s center. Therefore, the transfer velocity of irregularities in an arbitrary point of radio ray GL may be determined, if the components of the MIR station velocity perpendicular to ray GL are known: v_1 in the plane of Fig. 4.7.1 and v_2 in the perpendicular plane. The component v_1 in these experiments variated in the 1.4–2.4 km/s interval, and component v_2 changes from 1.3 km/s up to 4 km/s. During the first stage of data handling dependence $E(t)$ on time was transformed to dependence on minimal height of radio ray $E(H)$. Then at 50-70 km altitudes, value E_0 corresponding to the field strength in free space, was determined, and the normalization E / E_0 was provided. Altitude dependences of the dispersion and spectrum of the amplitude fluctuations σ_a^2 and $\Phi_a(F)$ were determined from analysis of δE fluctuations. The number of samples of the field strength E depended on the time lag Δt, which was used for determination of the dispersion σ_a^2. The time lag was selected from condition that for the time Δt the altitude H changes by 2 km, and the corresponding number of samples used for evaluation of σ^2 changed from 150 during the stratosphere sounding up to 2000 during the sounding of the lower part of troposphere. The spectrum of the field strength fluctuations $\Phi_a(F)$ was determined from results of measurements during the time Δt =2–7 s, corresponding to the 3–17 km altitude intervals. Obtained values of $\Phi_a(F)$ was smoothed by use of rectangular window having the constant quality-factor $F / \Delta F = 2$.

The typical altitude dependences of the field strength of centimeter waves are indicated in Fig. 4.7.2. Curve 1 and curve 2 characterize dependences $E(H)$, obtained, respectively, on 26.09.1996 and 29.09.1996 during immersion of the MIR station into the radio shadow. The median value of the field strength is diminishing and simultaneously the fluctuations caused by the atmosphere begin to appear. Dependences $E(H)$ in Fig. 4.7.2 present typical picture of radio wave fluctuations having strong spikes owing to the focusing effect and deep fadings. Because a variety of meteorological conditions dependences $E(H)$ were separated in two qualitatively different groups. In the first group the data, where the sharply expressed regular changes of $E(H)$ in the tropopause region at the altitude interval $H \approx 9$–13 km are observed, were included; curve 1 in Fig. 4.7.2 is relevant to this group Measurement sessions with heightened level of random fluctuations at the altitudes greater 13 km and absence of the strong focusing effect in the tropopause region were consolidated in the second group. Typical dependence $E(H)$ corresponding to this group is indicated in Fig. 4.7.2 by curve 2.

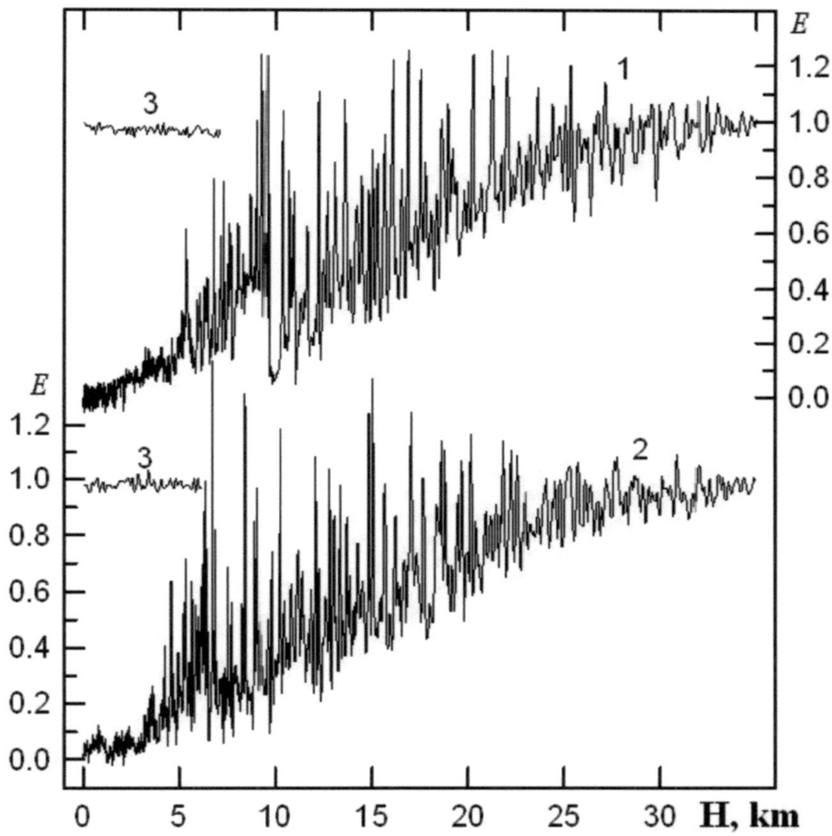

Fig. 4.7.2 Typical altitude dependences of the field strength of centimeter waves.

Quantitative characteristics of fluctuations are presented in Fig. 4.7.3 and 4.7.4 by values of the root mean square (rms) amplitude deviations σ_a for different altitudes H. Values σ_a obtained by means of analysis of dependences $E(H)$ for six measurement sessions relevant to the first group are shown in Fig. 4.7.3. One may see in Fig. 4.7.3 strong spread in values of σ_a at the 0.5–15 km altitudes. In the stratospheric region when $H > 15$ km values σ_a are not great and are diminishing when the height H increases. Fig. 4.7.4 describes dependence

$\sigma_a(H)$ obtained from data of six sessions, included in the second group. For this groups the regular growing of σ at the 35 km– 14 km altitudes *H*, a comparative constancy of the intensity fluctuations between 14 and 8 km and decreasing of σ_a between 8 km and 0.5 km are typical. In data, presented in Fig. 4.7.4 the statistical irregularities of the refractivity introduce the basic contributions into fluctuations, and influence of the layered structures seems to be insignificant. As follows from comparison of Fig. 4.7.3 and Fig. 4.7.4. that in the stratosphere the greater intensity

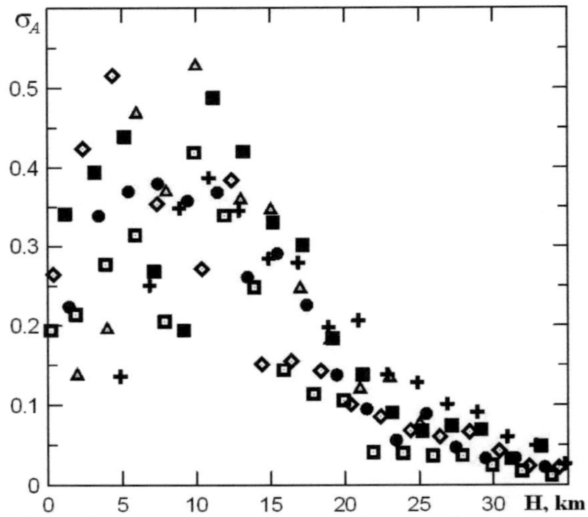

Fig. 4.7.3. Vertical profile of σ_a corresponding to the centimeter waves and the first data group.

of amplitude fluctuations is systematically observed in the second group. For example, values of σ_a at the altitude *H*=20 km are equal to 0.14 and 0.23, respectively, for the first and second group.

Let us analyze dependences $\sigma_a(H)$ and reveal indicators that may be used for estimation of the intensity of atmospheric turbulence. Two factors are affecting dependence of $\sigma_a(H)$: the background altitude profile of the refractivity $N(h)$ and refractive attenuation of radio waves *X*(*H*). Influence of the first factor may be accounted for by considering of the relative refractivity fluctuations $\delta N / N$. As follows from theory of wave propagation in the statistically inhomogeneous medium, value of σ_a is proportional to the RMS deviation σ_n of the refractivity fluctuations, therefore relative irregularity of medium may be characterized by the factor $\sigma_n/N \sim \sigma_a/N$. The second factor–influence of the refractive attenuation on the intensity of fluctuations may be accounted for by use of method described in [146]. As follows from data of this publication the refractive attenuation reduces the fluctuations in accordance with relationship

$$\sigma_a = \sigma_1 X^{5/6} \ . \qquad (4.7.30)$$

Here σ_1 is the RMS value of the amplitude fluctuations, which may be observed, if the actual atmosphere with strong refraction will be changed by the statistically

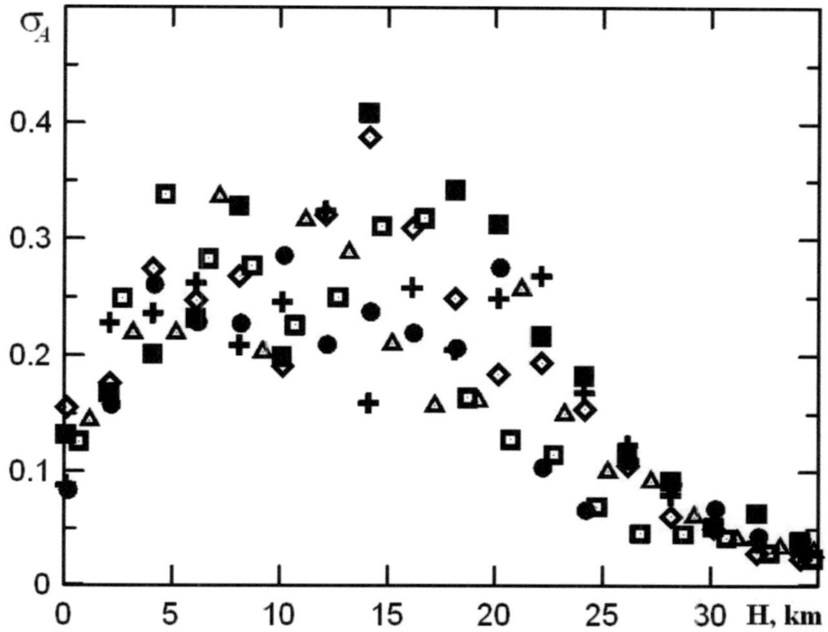

Fig. 4.7.4 Dependence $\sigma_a(H)$ obtained from data of six sessions, included in the second group.

inhomogeneous medium without refraction, X is the refractive attenuation of power of radio waves. Experimental dependences $\sigma_a(H)$, indicated in Fig. 4.7.3 and Fig. 4.7.4, validated this conclusion: really for the altitudes $H < 8$ km the decrease of intensity of the amplitude fluctuations is observed. Therefore, the altitude dependence in the form

$$U(h) = \sigma_a N^{-1} X^{-5/6} \tag{4.7.31}$$

may be considered as an experimental characteristic of intensity of atmospheric turbulence at different altitudes. For numerical determination of the characteristic $U(h)$ the experimental dependences $\sigma_a(H)$, $X(H)$ and known altitude profile $N(H)$ may be applied. In Fig. 4.7.5 the results of determining of dependence $U(h)$ are presented.

As follows from comparison of curves 1 and 2 in Fig. 4.7.5 the characteristic U of the first and the second groups is clearly different. The function U corresponding to the second group has heightened value in stratosphere at altitudes 20-30 km. In this region the function U depends weakly on the height h. For the altitudes below 20 km the monotonic decrease of U is observed with decreasing height. Quick decrease of U in the 30–23 km height interval and constancy of U at the altitudes h=13-22 km is a feature of the first group. For the altitudes below 13 km in the function U relevant to the first group has greater value than the same one corresponding to the second group. This is connected with essential contribution of layered structures in the experimental values σ_a relevant to the first group. Therefore, the characteristics $\sigma_a(h)$ and U for $h < 13$ km do not give objective information on the atmospheric turbulence. Spectrum of the amplitude fluctuations gives additional information concerning the atmospheric irregularities. Examples of the amplitude

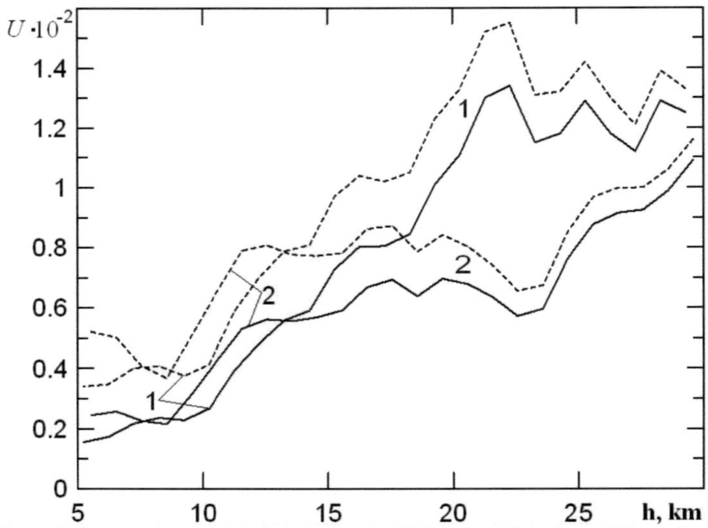

Fig. 4.7.5 Results of determining of dependence *U*(*h*) for the first (2) and second (1) data groups.

fluctuations spectra observed during radio sounding of two regions of the atmosphere are shown in Fig. 4.7.6. Spectrum A was obtained on 27.05.1998 above Kazachstan, and spectrum B was observed on 08.07.1998 above the Indian Ocean. Vertical bars in Fig. 4.7.6 indicated the 90% confidence interval. Spectrum $\Phi_a(F)$ is typical for the amplitude fluctuations and is characterized by: (1) nearly constant spectral density in the low frequency region, (2) the inflection frequency of spectrum F_0, and (3) power law for changes of the spectral density $\Phi \sim F^{-m}$ in the interval $F > F_0$. Let us consider the changes spectral indexes m and q in the stratosphere and troposphere.

Values m and q were determined from results of spectra analysis, obtained during sounding of the atmosphere in 214 regions. In Fig. 4.7.7 the average values of the index m (left vertical axis) and spectral index q (right vertical axis) are indicated by points for different heights H. Each point in Fig. 4.7.7 is obtained in results of averaging of 15-18 spectra, vertical bars characterize the rms deviation of parameter m, and the brocken dotted line describes the averaged experimental dependence $m(H)$. One may select in these dependences three intervals: in the troposphere when $H_0 \leq 7$ km the index m has small changes and its median value is equal to $m = 2.5 \pm 0.2$. This value corresponds well to the index of the Kolmogorov' spectrum $q = m + 1 = 11/3$. In the stratosphere when $H =$ 15-30 km the experimental median value m =3.5 $\pm$ 0.2. It is practically independent on the height, the corresponding spectral index $q = 4.5 \pm 0.2$ is essentially distinct as compared with the theoretical value $q = 11/3$. At altitudes 7– 15 km the experimental values m change in a broad interval that connected, mainly, with different height of the tropopause in the sounding regions. Linear dependence $m(H)$, indicated in Fig. 4.7.7 for this height interval, is very conditional. Difference in meteorological conditions in the investigated regions determines random character of the index m variations in these altitudes.

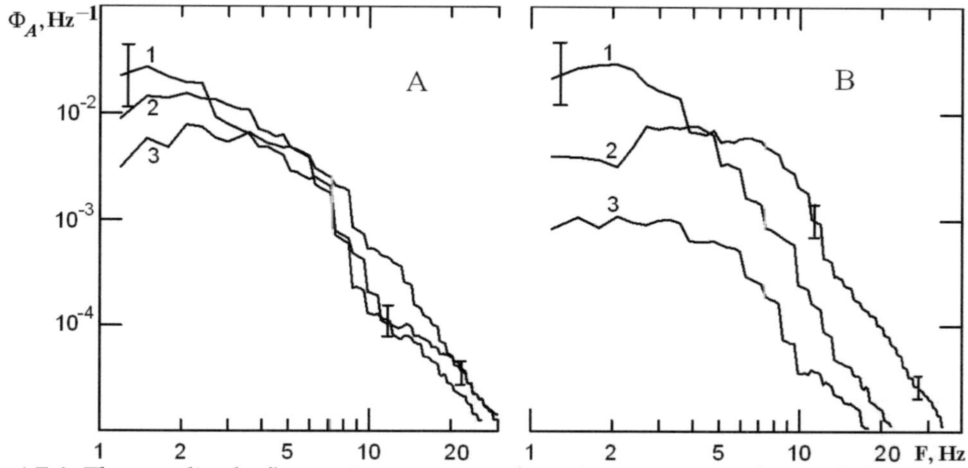

Fig. 4.7.6. The amplitude fluctuations spectra of centimeter waves observed during radio occultation of two regions of the atmosphere. Spectra 1, 2 and 3 are relevant to the altitudes H=7, 22 and 26km.

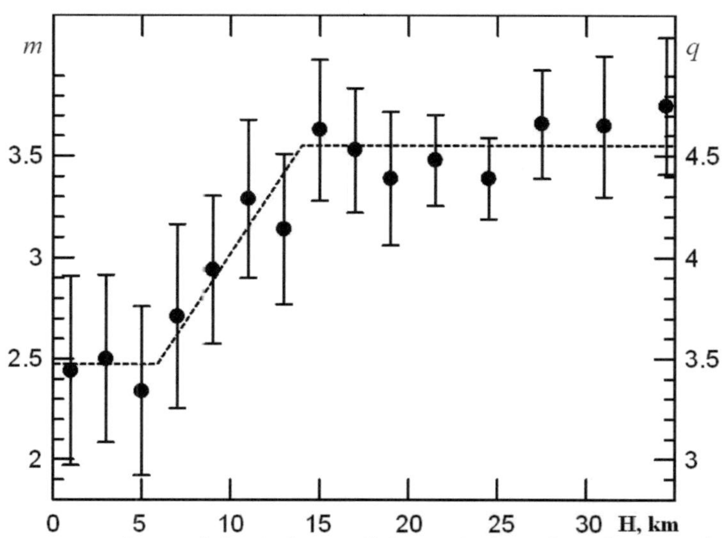

Fig. 4.7.7 The average values of the index m (left vertical axis) and spectral index q (right vertical axis) for different heights H.

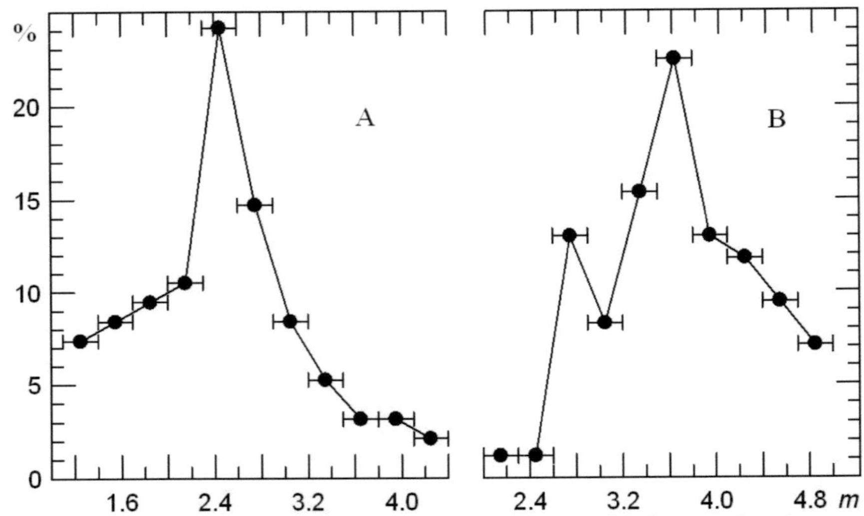

Fig. 4.7.8 Empirical distribution of the probability density of values m for the troposphere (A) and stratosphere (B) measured in the centimeter wave's range.

Let us consider the empirical distribution of the probability density of values m, presented in Fig. 4.7.8. Values m are indicated along the horizontal axis, and the frequency of observation expressed in percents and corresponding to values m in the interval $\Delta m = 0.3$– along vertical axis. The distribution m in the troposphere is indicated in Fig. 4.7.8A, and the distribution m in the stratosphere is shown in Fig. 4.7.8B. It may be seen, that the distributions in the troposphere and stratosphere are essentially distinct. In the troposphere in the 50% cases $m = 2.5 \pm 0.5$ and the corresponding to this value spectral index q is equal to $q = m + 1 = 3.5 \pm 0.5$, i.e. it is near to value 11/3. characterizing the classic turbulence. In the stratosphere in 50% the cases $m = 3.6 \pm 0.5$; an additional maximum with $m = 2.7$ is observed in the distribution. Therefore, in the stratosphere only 13% realizations of the observed values of the index m corresponded to the Kolmogorov' spectrum, that reveals the different from turbulence mechanism of irregularities formation.

Deviation of irregularities structure in the stratosphere from classical turbulence requires accounting for the anisotropy of irregularities in analysis of the amplitude fluctuations. The spectrum $\Phi_a(F)$ for stratospheric region will be analyzed in more detail with usage of results obtained in papers [29, 143–145]. Let us consider at first parameter, characterizing spectrum of the amplitude fluctuations: the inflection frequency F_0. The inflection frequency F_0 has been determined from location of maximum of the dimensionless value $F\Phi_a(F)$ on the frequency axis. This location has beedn obtained from analysis of 214 spectra, registered during radio occultation of the atmosphere at 0.3– 30 km altitudes interval. Variations F_0 were in the 1.2 - 1.5 Hz interval when the intersection velocity v in point T changes from 2.5 up to 5.2 km/s and the immersion angle α was in the 23° - 71° range. Evaluation from formula (4.7.26) gives for changes F_0 the 1.4 –10 Hz range. Difference in the experimental and theoretical values F_0 may be connected with influence of anisotropic irregularities. The anisotropy is characterized by the coefficient η, determined as a ratio of the horizontal scale of irregularities to vertical scale. Changes of F_0 occur also because of decrease of vertical size of the first Fresnel' zone and the immersion velocity owing to refraction. Characteristic changes of the frequency F_0 as function of height depend also on the immersion angle α (Fig. 4.7.1). Asymptotic formula for F_0 has been obtained in [140] for two limiting cases: vertical immersion in the atmosphere ($\alpha=0$), when $F_0 \sim X v_1$, and the horizontal immersion ($\alpha=90^\circ$), when $F_0 \sim v_2$. When $\alpha \neq 0$ the anisotropy of irregularities may cause the decrease of the frequency F_0, however in this case the form of the spectral density of fluctuations in the frequency band $F > F_0$ does not change.

Let us discuss connection between the spectrum $\Phi_a(F)$ and spatial spectrum of the refractivity fluctuations Φ_n. The frequency spectrum $\Phi_a(F)$ may be transformed in 1-dimensional spatial spectrum of the refractivity fluctuations $v_n(K)$ by use of the "frozen" hypothesis. The 1-dimensional spatial spectrum of the refractivity fluctuations may be presented in the form

$$V_n(K)=(4\pi)^{-1}\ v\ \Phi_a(K\ v/2\pi), \qquad (4.7.32)$$

where $K=2\pi/\Lambda=2\pi F/v$ is the wave number, Λ is the scale of irregularities, $\Phi_a(F)$ is the frequency spectrum of the amplitude fluctuations. If irregularities of the refractivity are statistically isotropic, then the spatial spectrum $V_n(K)$ should not depend on the angle α. However dependence of the spectrum $V_n(K)$ on α is observed in experimental data.

It was noted in [146, 147] that in the case of highly stretched along the horizontal direction irregularities, i.e. in the case of layered structures, the frequency spectrum of the amplitude fluctuations $\Phi_a(F)$ should be determined mainly by the vertical component of velocity v_1 and the frequency spectra should be similar after transformation to their vertical wave number $\Lambda=2\pi F/v_1$. 1-dimensional spatial spectrum along vertical is determined in this case from formula (4.7.22) when $K=K_y$, $v=Xv_1$ where v_1 is vertical velocity of point T with accounting for the refraction effect. In Fig. 4.7.9 examples of the experimental spectrum $v_n(K_y)$, obtained from measurements above the Indian Ocean (black signs) and above Turkmenia (bright signs) are shown. Values of K_y and vertical scale $\Lambda=2\pi/K_v$, and the spectral density V_n are marked, respectively, on the horizontal and vertical axis. Squares in Fig. 4.7.9 indicated the spectrum, observed at the altitude H=20 km, triangles- at H=25 km, and circles and rhombuses- at H=30 km . Measurements were conducted above the Indian Ocean when the angle α was equal to $\alpha=28^\circ$, measurements above Turkmenia correspond to $\alpha=71^\circ$. Despite different meteorological conditions in these regions and distinctive trajectories of immersion into the atmosphere, the spectra V_n in Fig. 4.7.9 demonstrate good reproducibility, that validates the locally isotropical irregularities assumption. In the stratosphere at the altitudes 15- 35 km the spectra V_n have maximum spectral density at the scales Λ=300–800 m. At the scales $\Lambda \leq 200$ m the spectral density

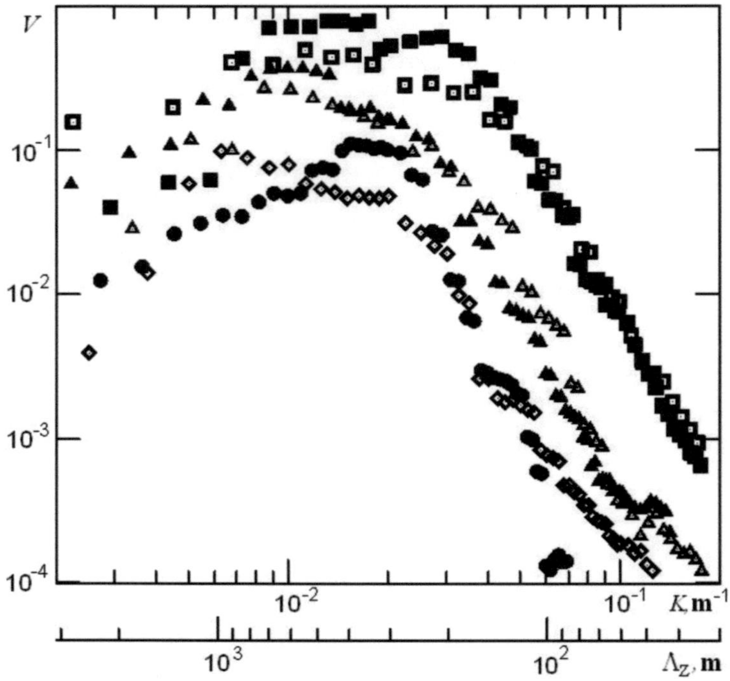

Fig. 4.7.9 Examples of the experimental spectrum $v_n(K_y)$ as function of vertical wave number.

V_n is diminishing as $K_y^{-4.5}$ up to scale $\Lambda \approx 40$ m, i.e. the spatial spectrum of irregularities in the stratosphere is different as compared with the Kolmogorov' spectrum. Therefore, from results of analysis of the amplitude fluctuations spectra with different immersion angles one may reveal features of the mechanisms responsible for irregularities formation. In [146,147] it was noted, that the frequency spectra of the intensity fluctuations of the stars' light registered during immersion in the earth's atmosphere contain information about the structure of small-scale irregularities of the atmospheric density and processes, taking part in their formation. In the optical range such data may be obtained only for the large altitudes in the stratosphere.

Measurement of the fluctuations of radio waves in the centimeter range in the communication links satellite-to-satellite are more appropriate for studying the structure of atmospheric irregularities. In [145] results of the atmospheric sounding in radio link MIR station–geostationary satellites were used for detecting in the stratosphere saturated internal gravity waves in the case of the developed isotropic turbulence. Presentation of the 3-dimensional spectrum of the refractivity fluctuations in a general form, admitting transfer as to the isotropic turbulence, and to the anisotropic model of the irregularities spectrum, allows explaining the features in the experimental frequency spectrum of the amplitude fluctuations $\Phi_a(F)$. It is turned out, that presentation of the inhomogeneous atmosphere by an equivalent phase screen and usage of the known from paper [148] form of the fluctuations spectrum, caused by saturated internal gravity waves, admits to compile a full equation system, connecting the experimental frequency spectrum $\Phi_a(F)$ with 3-

dimensional spectrum $\Phi_n(K)$. By setting parameters into the spectrum $\Phi_n(K)$ for selected type of irregularities one may recalculate the spectra of the amplitude fluctuations $\Phi_a(F)$, coinciding with experimental data, and determine the intensity of irregularities, spectral index q and estimate the quality of the applied irregularities model. Analysis indicated that for the effective separation of the contributions of the isotropic and anisotropic irregularities it is necessary to provide measurements in the case when the plane of immersion or egression of the radio ray GL inclined, if possible, under the large angle α relative to the plane, passing through the points G, L and the earth' s center. This can be explained because at small angles α the ray line GL intersects the anisotropic irregularities mainly in vertical direction and, as it seen in Fig. 4.7.9, the spectrum of the amplitude fluctuations become similar after their recalculation to vertical wave numbers. If the anisotropy coefficient η >> 1, this phenomenon is realized even for very «inclined» occultation and the frequency of the amplitude fluctuations is determined by vertical component of the point T velocity $v_y = v\cos\alpha$. With increasing of angle α, the velocity v_y is diminishing and this leads to growing of the contributions of the anisotropic irregularities in the low frequency region of spectrum $\Phi_a(F)$. If the full velocity v is growing, contributions of isotropic irregularities is diminishing in the low frequency region spectrum and simultaneously the inflective frequency F_0 is displacing into the high frequency range of spectrum. Therefore, separation of the anisotropic and isotropic components of the amplitude fluctuations is possible when the immersion angle α is changing. Conducted in [145] analysis of experimental spectra indicated, that at frequency F_0 ~ 3.5-4 Hz contributions of anisotropic irregularities of density prevail, and at the frequency F_0 ~ 13 Hz – isotropic irregularities are significant. Significant separation of the considered spectra on frequency is essential for their detecting. It is turned out that in the stratosphere the amplitude fluctuations are mainly determined by anisotropic irregularities introducing main contribution in the dispersion σ_a^2, the turbulence contribution is small. This is well agree with data shown in Fig. 4.7.7 and Fig. 4.7.8. As follows from Fig. 4.7.7 and Fig. 4.7.8 one may observe satisfactory correspondence of the RO and radiosounders measurements with model of the saturated internal gravity waves, considered in [149].

Let us consider, following to [150–152], features of the amplitude and phase fluctuations of the decimeter radio waves. Investigations of the amplitude fluctuations were accomplished with usage of the decimeter radio waves at $\lambda = 19$ cm, when the satellite Microlab registered signals of GPS satellites. Because movement of the satellites the minimal altitude of radio ray H is diminishing, and changes of the amplitude and phase radio waves, propagating through the atmosphere were registered. Variations of the amplitude E and phase φ of RO signal have two components: regular changes, caused by the background profile of the refractivity $N(h)$, and the statistical fluctuations δE and $\delta\varphi$, connected with irregularities of the refractivity. For determination of the statistical characteristics of fluctuations δE and $\delta\varphi$ the regular changes of the phase and amplitude were found by means of the second power polynomial approximation on the short intervals of time Δt; then from dependences $E(t)$ and $\varphi(t)$ changes of the phase and amplitude, corresponding to the regular trends, were subtracted and the random functions δE and $\delta\varphi$ with zero mean values were obtained. Experimental dependences δE and $\delta\varphi$ were analyzed by usage of the standard procedure of statistical data handling. As a result for each RO session several values of the dispersion of the amplitude fluctuations σ_E^2 and phase σ_φ^2 were found. During RO experiments radio waves are propagating also through the ionosphere. To exclude the ionospheric influence and technical factors the fluctuation's characteristics of the radio signals at the altitude *H*=40-60 km were determined, where atmospheric influence may be neglected. Exception of small influence of the ionospheric irregularities and technical noise on the phase and amplitude fluctuations was conducted by a usual procedure applicable for statistically independent random processes, assuming the dispersions and spectral densities may be subtracted.

To elucidate the laws of the decimeter waves fluctuations the data of the 49 RO sessions accomplished in period February 2– 19, 1997 in the regions with the longitude 120°-153°E and latitude 20°-60°N were used. The distances L_l, L_g from the receiving Microlab satellite and transmitting GPS satellite to the radio ray perigee T (Fig. 4.7.1) were equal to $L_l = 3300$ km and $L_g = 25{,}600$ km, respectively. The velocity of point T has two components: the vertical $v_1 = dH / dt$ and horizontal v_2, which is perpendicular to the plane of Fig. 4.7.1. Only experimental data relevant to the nearly vertical immersion (or egression) were analyzed with $v_2 < v_1$. Dispersion of the amplitude fluctuations was determined at the time lag $\Delta t = 4$ s. During this time the altitude *H* diminished by 7-10 km in the stratosphere and by 1.8-2.6 km in the troposphere, therefore during each RO session five-seven σ_a values were obtained. In Fig. 4.7.10 results of determination of the RMS deviation σ_a of amplitude of decimeter radio waves are presented for different altitudes *H*.

The spread of experimental values σ_a in this Figure is caused by variability of the meteorological conditions, and averaged dependence σ_a on *H* is indicated by the solid broken curve. Reliable registration of the atmospheric amplitude fluctuations is beginning in the stratospheric region at the altitudes *H* = 23-26 km, with average $\sigma_a = 0.05$. At the altitudes *H* = 13–17 km the fluctuations intensity remains nearly constant with average value $\sigma_a = 0.15$. In the troposphere at the altitudes *H* = 3-6 km a strong spread of values σ_a with average $\sigma_a = 0.31$ is observed. In Fig. 4.7.11 dependence of the RMS phase deviation on the height is indicated. Points correspond to values σ_φ (expressed in radians) for different RO sessions, and the solid broken curve corresponds to average experimental dependence. Analysis of the phase fluctuations indicated, that determination of σ_φ is reliable starting from the heights 23-26 km with the average value σ_φ equal to 0.18 radian. The monotonic increase of the phase fluctuations is observed when *H* is decreasing. At the altitudes 13-17 km value σ_φ in average is equal to $\sigma_\varphi = 0.53$. In the troposphere the large variability of the phase fluctuations is observed: the registered in different RO sessions values σ_φ are changing from 0.5 up to 2.6 radian with average values 1.2 in the height interval *H* = 3-6 km.

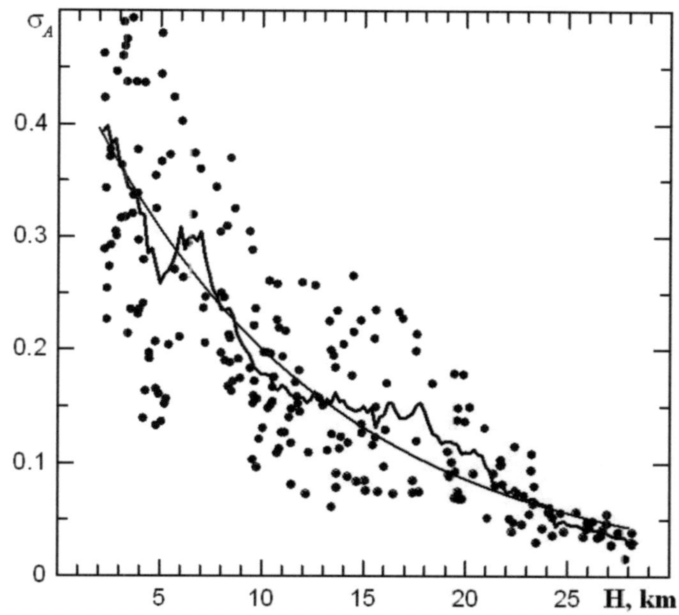

Fig. 4.7.10. Vertical profile of the RMS deviation σ_a of the amplitude of the decimeter radio waves.

From comparison of Fig. 4.7.10 and 4.7.11 it follows, that dependences $\sigma_a(H)$ and $\sigma_\varphi(H)$ are similar. In Fig. 4.7.12 the ratio of the average values $\sigma_\varphi / \sigma_a$ is presented as function of the altitude H. As follows from Fig. 4.7.12 the ratio $\sigma_\varphi / \sigma_a$ is equal to $\sigma_\varphi / \sigma_a = 3.4 \pm 0.6$ and weakly depends on H.

Let us consider the spectrum of the amplitude and phase fluctuations. Spectra of the amplitude fluctuations were determined in the time interval $\Delta t = 5$ s, corresponding changes of the minimal height of radio ray were in the intervals $\Delta H = 8\text{-}12$ km in the stratosphere and $\Delta H = 2.2\text{-}3.0$ km in the troposphere. In Fig. 4.7.13 the typical spectra of the amplitude fluctuations $\Phi_a(F)$ are presented for the three RO sessions. The spectrum «A» is obtained for the troposheric, and spectrum «B»– for the stratospheric region, respectively, curves 1, 2, and 3 correspond to different RO sessions, where the significantly distinguished levels of the fluctuations were observed. From Fig. 4.7.13 it follows, that at frequency $F < 2$ Hz the spectral density $\Phi_a(F)$ is weakly depending on the frequency F, and when $F > 2$ Hz $\Phi_a(F)$ is quickly diminishing with increasing of the frequency. For $F > 2$ Hz the experimental spectrum is well approximated by dependence $\Phi_a \sim F^{-m}$ and the spectral index m may be determined from the experimental data. In Fig. 4.7.14, the histograms indicating the distribution of the index m values, obtained from amplitude data, are shown. From Fig. 4.7.14, it follows, that in the troposphere in the 75% cases the spectral index m values are located in the interval 2.2-3.4 with the average values equal to 2.7±0.6, and in the stratospheric region the spectral index m changes from 2 up to 4. From comparison of the

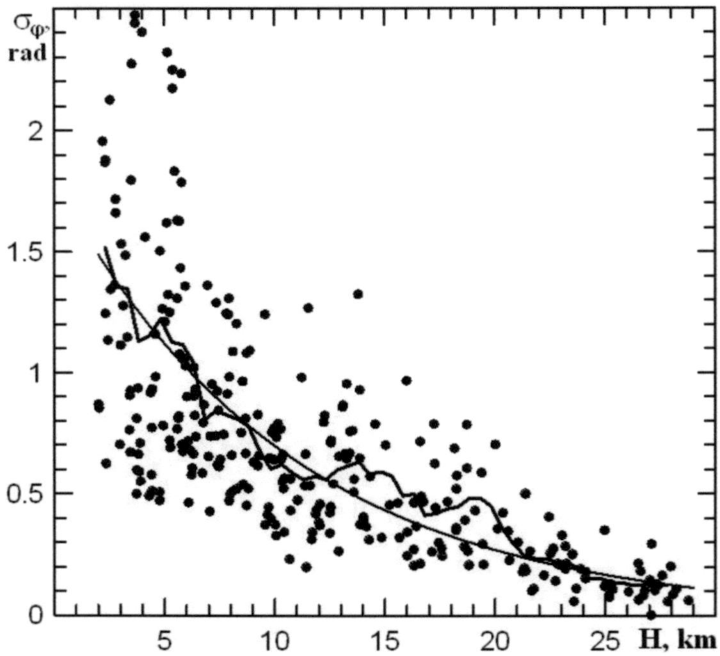

Fig. 4.7.11. Dependence of an RMS phase deviation of the decimeter waves on the height *H*.

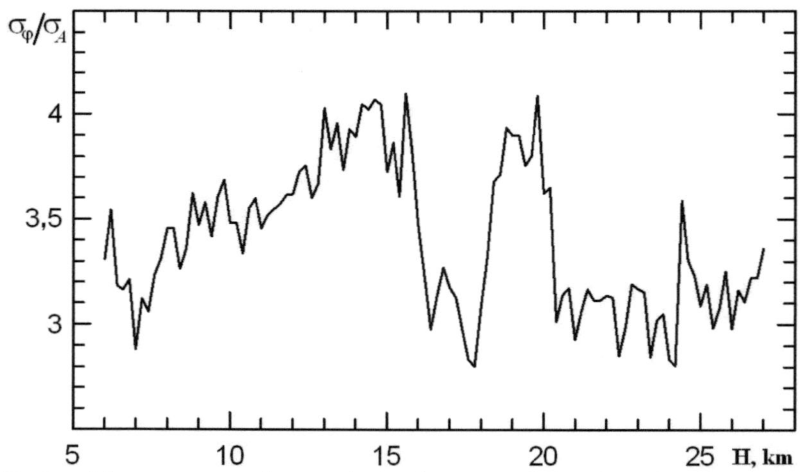

Fig. 4.7.12. Ratio of the average values $\sigma_\varphi / \sigma_a$ as function of the altitude *H*.

index m distribution found from the amplitude fluctuations of the centimeter (Fig. 4.7.8) and decimeter waves (Fig. 4.7.14), it follows the good correspondence of these data, that indicates the reliability of determination of the spectral index from the refractivity fluctuations q . Let us adduce the results of analysis of the experimental spectra fluctuations of the eikonal $\Phi_\psi(F)$. The spectra $\Phi_\psi(F)$ were determined in the time interval $\Delta t = 5$ s; in

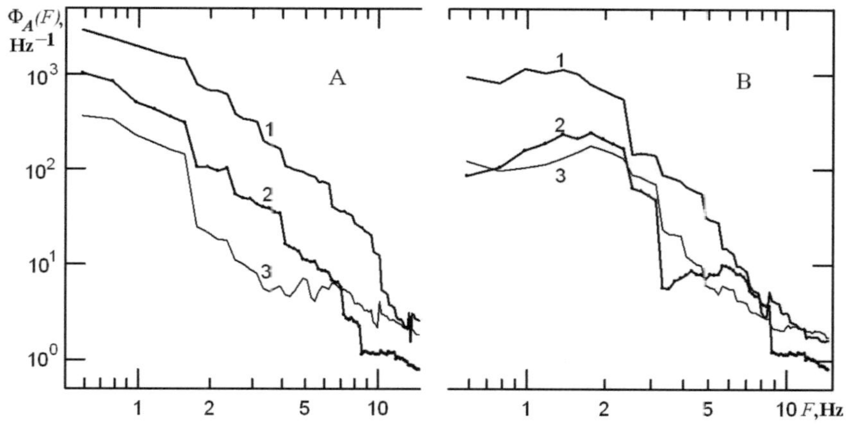

Fig. 4.7.13. The typical spectra of the amplitude fluctuations $\Phi_a(F)$ of the decimeter radio waves.

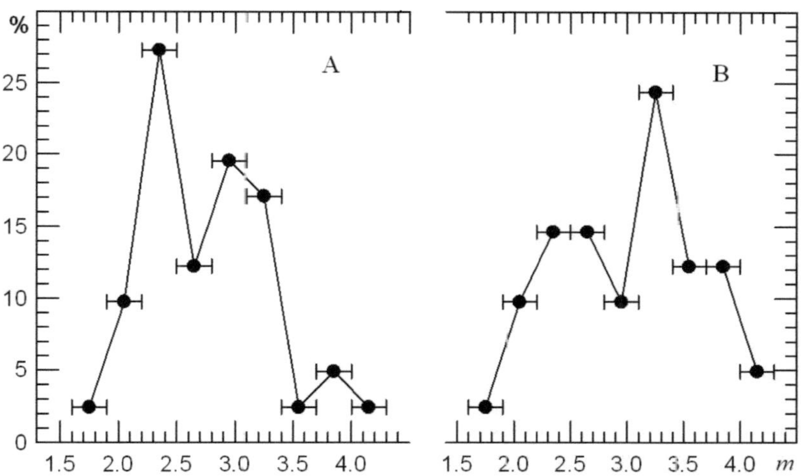

Fig. 4.7.14. Histograms indicating the distribution of the index m values, obtained from amplitude data in the decimeter range for the troposphere (A) and stratosphere (B).

the stratospheric and tropospheric regions the data were used, corresponding to the altitudes H 15–24 km, and 4–7 km, respectively. In Fig. 4.7.15 the examples of the experimental spectra of the eikonal fluctuations for the troposheric and stratospheric regions are shown. The experimental spectra Φ_ψ are also well describing by the power dependence $\Phi_\psi \sim F^{-m}$. It is turned out, that in the troposphere the average value m is equal to 2.2±0.4. In the stratosphere $m = 2.6\pm0.5$; therefore, according to (4.7.22), for the stratosphere $q = 3.6\pm0.5$, and for the troposphere $q = 3.2\pm0.4$. This corresponds to the Kolmogorov' spectrum with $q = 11/3$. As follows from Fig. 4.7.13 and 4.6.15 the form of the spectra Φ_a and Φ_ψ is successfully obtained for the 0.7– 20 Hz interval of the frequency F. This corresponds to the irregularities scale Λ in the interval 0.1– 2 km when the velocity changes in the interval 1.9-2.6 km/s.

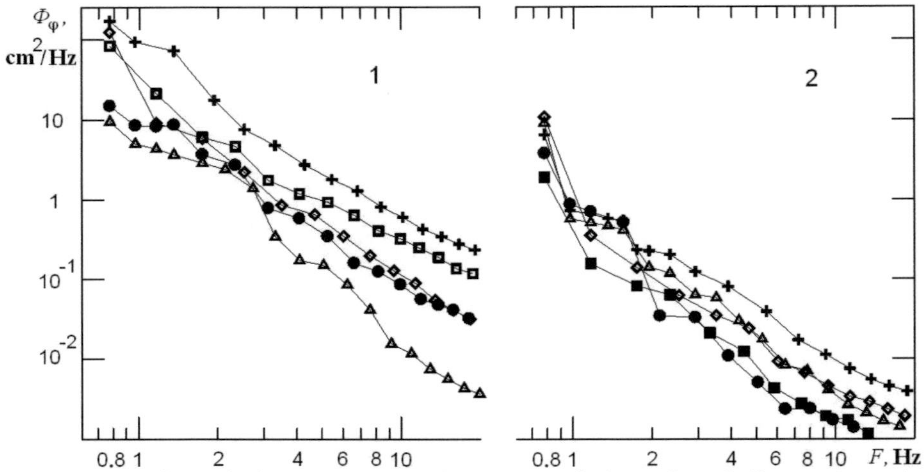

Fig. 4.7.15 Examples of the experimental spectra of the eikonal fluctuations for the troposheric and stratospheric regions.

Let us compare experimental data containing fluctuations of the amplitude and phase with theoretical results. For determination of connection between the amplitude fluctuations σ_a^2 and phase σ_φ^2 and the altitude dependence of the dispersion of the refractivity fluctuations $\sigma_{\mathrm{n}}^{\;2}(h)$ the approximated relationship (4.7.27) and formulas (4.7.28), (4.7.29) may be applied to analysis. The value $\sigma_\varphi / \sigma_a$ was successfully determined from experimental data; therefore the ratio $\sigma_\varphi / \sigma_a$ should be formed theoretically from Eqs. (4.7.28) and (4.7.29):

$$\left(\frac{\sigma_\varphi}{\sigma_{\grave{a}}}\right)^2 = \mathrm{k}^{5/6} \Lambda_{\mathrm{m}}^{\;5/3} \left(\frac{\mathrm{L}_l + \mathrm{L}_g}{\mathrm{L}_l \mathrm{L}_g}\right)^{5/6} . \tag{4.7.33}$$

It follows from (4.7.33), that the ratio $\sigma_\varphi / \sigma_a$ is determined by the wavelength, the distances L_l, L_g, the external scale of irregularities Λ_{m}, and is independent on the height *H*. Expression (4.7.33) allows determining the external scale of irregularities from known ratio $\sigma_\varphi / \sigma_a$. After substitution in Eq. (4.7.33) the experimental value $\sigma_\varphi / \sigma_a$ = 3.4, the magnitudes of the distances L_l, L_g and wavelength λ, the average value Λ_{m} may be found equal to $\Lambda_{\mathrm{m}} \approx 1.4$ km. After comparison of the experimental dependences $\sigma_a(H)$ and $\sigma_\varphi(H)$, presented in Fig. 4.7.10 and 4.7.11 by the broken curves, with results of evaluation of these dependence from the Eqs. (4.7.28) and (4.7.29), indicated in Figures 4.6.10 and 4.6.11 by smooth curves, one may found the parameters β and σ_0. Experimental data and results of evaluation are coinciding well, if one prescribes values 0.1 km^{-1} and $2{\cdot}10^{-6}$ to the parameters β and σ_0, respectively. The RMS deviations of refractivity fluctuations σ_{n} corresponding to these values β and σ_0 are equal to $1.2{\cdot}10^{-6}$ and $0.4{\cdot}10^{-6}$ at the altitudes *h* 5 and 15 km, respectively. One may estimate the width of an equivalent irregular «layer» $(\pi a\beta^{-1})^{1/2}$; after substituting the found value β one may conclude that value of the «layer» width is equal nearly to 400 km.

Characteristics of the amplitude and phase fluctuations of the centimeter and decimeter radio waves in the communication satellite-to-satellite links are satisfactorily described by results of radio wave propagation theory in statistically inhomogeneous medium. Comparison of experimental data with theoretical results allows determining the index of the spatial spectrum of atmospheric irregularities q, estimating the external scale of irregularities Λ_m and finding averaged dependence of the dispersion of refractivity fluctuations as function of the height $\sigma_n^2(h)$. The index of spatial spectrum of the atmospheric refractivity irregularities q may be determined independently from the frequency spectra of the amplitude and phase fluctuations that essentially heighten the trustworthiness of the q determination. The found from the amplitude and phase data values q correspond in the troposphere to the Kolmogorov' spectrum of developed turbulence and one may consider as a validated fact, that the amplitude and phase fluctuations at the altitudes 2-8 km are really caused by atmospheric turbulence having the index magnitude $q = 11/3$. In the troposphere the temperature is decreasing with the height increasing, the atmosphere is instable and therefore intensive turbulence develops easily. In the stratosphere the temperature weakly increases with height or the isothermal region develops, therefore the atmosphere is stable. In the stable atmosphere layered and wave structure develop, and classical turbulence become apparent as a small factor. Therefore the registered values of the index m and, hence, the index of the spatial spectrum of atmospheric irregularities q change in the large interval. Small-scale atmospheric irregularities may be characterized by the intensity of the amplitude and phase fluctuations, i.e. by values σ_φ and σ_a. The parameters σ_φ and σ_a are theoretically connected with the dispersion of the refractivity fluctuations σ_n^2.

Determination of the altitude dependences $\sigma_n^2(h)$ may be reliably fulfilled only by averaging of many RO sessions data. In a separate RO session one may observe the intensive changes of σ_φ and σ_a, hence determination of σ_n become difficult. Therefore the intensity of the small-scale atmospheric irregularities may be better characterized by the dispersions of the phase σ_φ^2 and amplitude σ_a^2 fluctuations without recalculation of these values into $\sigma_a(H)$. To heighten reliability of obtained information about the intensity of atmospheric irregularities it is necessary to use results of the simultaneous amplitude and phase measurements. One may consider the value

$$\sigma = \frac{1}{2}(\sigma_\varphi + 3.4\sigma_a).$$

as the conditional characteristic of the intensity of the atmospheric irregularities, obtained from RO monitoring of the atmosphere. In the troposphere, in the most cases $q = 11/3$, and the parameter σ describes the intensity of turbulence. In the stratosphere, where the wave and layered structure are observed, σ may not be used for the estimation of the turbulence intensity. Analysis of the frequency spectrum of amplitude and phase fluctuations allows revealing the relative contributions of atmospheric waves and Kolmogorov' turbulence of small-scale stratospheric irregularities. By applying the described results of investigation of radio wave fluctuations and by use of formula (4.7.17), one may estimate the temperature fluctuations δT. As follows from considered data the RMS deviation of temperature σ_t at the altitudes 4-8 km is equal to 1.2-1.7 K, and at the altitude 20 km $\sigma_t \approx 1.8$-2.4 K. The fluctuations δT characterize the maximal accuracy in determination of altitude profile of temperature $T(h)$ by RO method, when other interfering factors are removed. Comparison of dependences $T(h)$, determined by RO method and other tools is conducted in [105]. It is shown that errors of temperature determination are nearly equal to the estimated values σ_t. For studying small-scale irregularities in the stratosphere in papers [146, 147, 153, 154] the space borne observations of effect of star brightness scintillations during immersion or egression were used. Comparison of the amplitude fluctuations of radio and optical waves during occultation sounding of the stratosphere was conducted in [155]. This comparison indicated satisfactory correspondence with considered above experimental data.

5. RO Monitoring Of The Lower Ionosphere

5.1. Influence of the lower ionosphere on the amplitude and phase of RO signal

Physical changes in the near-earth space environment in response to variations in solar radiation, solar plasma ejection, and the electromagnetic status of the interplanetary medium produce disturbances in the ionosphere. The disturbed ionosphere changes the amplitude and phase of RO signal. To lowest order, changes in the total electron content (TEC) along the signal path contribute to the phase path excess. For an undisturbed ionosphere, where the electron density does not vary significantly over the short- scale lengths, this is the only effect that the ionosphere has on the RO signals. For undisturbed conditions the tangent points in the ionosphere are absent during motion of the ray perigee in the atmosphere and the ionospheric influence may be described as a slow change (appeared as linear or parabolic trend) in the phase path excess without noticeable variations in the amplitude of RO signal. As shown in [73], this trend can be extrapolated to the atmospheric part of RO measurement and then subtracted from the phase path excess to measure the atmospheric contribution. Thus, in the quiet ionospheric conditions the ionospheric correction can be introduced independently to the phase path excess at both RO signals. In the case of the CHAMP radio occultation experiment the quiet ionospheric conditions have come to light in the form of smooth dependence of the amplitude on the height interval 40–100 km with small dispersion of the amplitude variations σ^2

$$\sigma = [< (A - < A >)^2 >]^{1/2} / < A >, \tag{5.1.1}$$

where $< A >$ is the average amplitude of RO signal corresponding to propagation in the free space. An example of quiet ionospheric conditions observed during CHAMP RO event is shown in Fig. 5.1.1. The amplitude curve has low amplitude variations RMS=0.0136. The phase data (curve 1 and 2 are relevant to frequencies L1 and L2) are obtained after

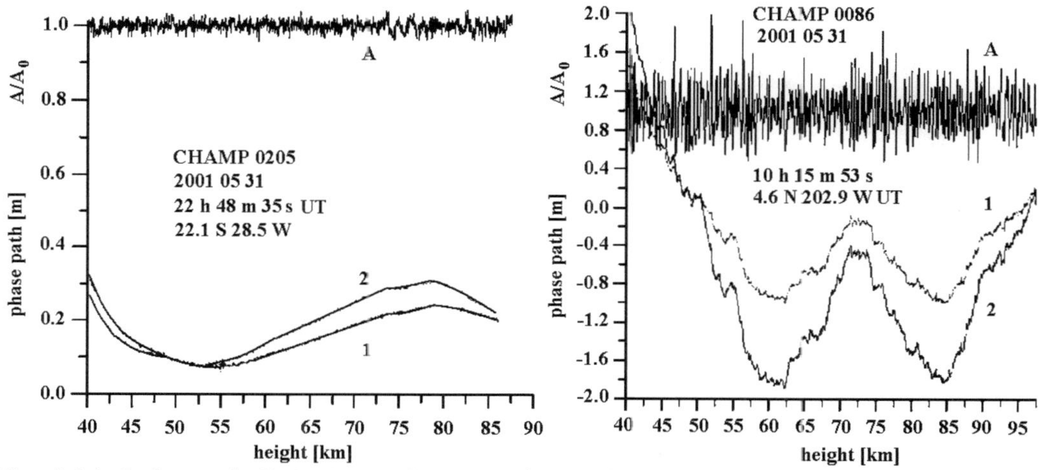

Fig. 5.1.1. Left panel. Quiet ionospheric conditions (rms=0.0136), CHAMP RO event No. 0205, May, 31, 2001. Insert indicates the date, time of event, and coordinates of the ray perigee. Right panel. Noisy CHAMP event with high amplitude variations (RMS=0.2191) in the equatorial region at the local midnight.

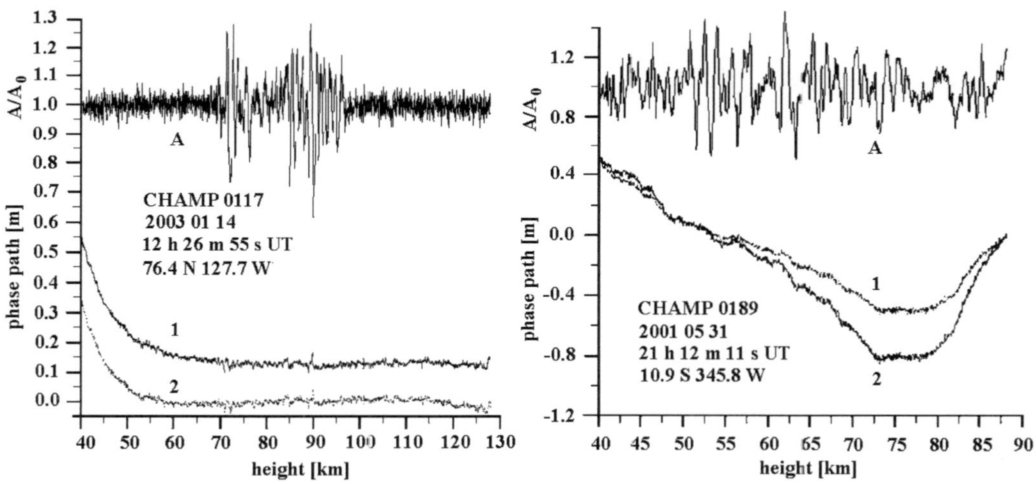

Fig. 5.1.2. Regular variations in the amplitude and phase can be connected with waves in the electron density distribution for polar (left) and equatorial evening ionosphere (right).

a linear trend. The phase path excess at both frequencies reveals very slow changes connected with slow altering of the TEC along the ray path in the ionosphere. An example of CHAMP event with high amplitude variations (rms = 0.2191) is shown in Fig. 5.1.1 (right panel). The noisy contribution may correspond to a fountain phenomenon in the ionosphere usually observed after sunset during the time of high solar activity. This event can be classified as a typical case, relevant to the noisy ionospheric contribution caused by turbulence in the equatorial region after sunset at 21 h local time (LT). The main part of the amplitude variations is limited to the interval $0.6 \leq A \leq 1.3$ with some spikes up to level 1.9-

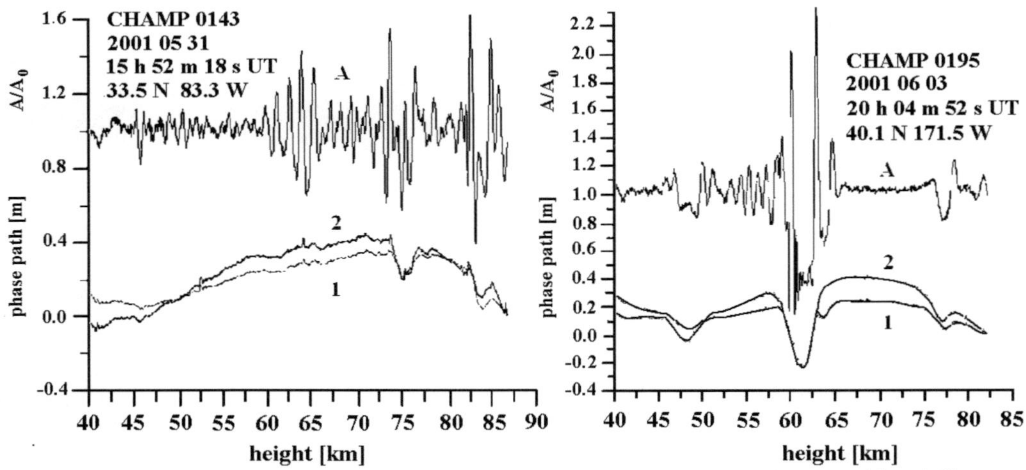

Fig. 5.1.3. Diffraction event in the morning ionosphere (08 h 40 m LT) caused by diffraction of radio waves on the sharp vertical gradients of the electron density in the sporadic E- layer of the ionospherec(right). Regular amplitude and phase variations connected with the inclined sporadic E-layers (morning ionosphere, 09 h 30 m LT)(left).

2. Phase variations (after subtracting of the linear trend) are also high and indicated sharp changes in the TEC corresponding to the moving bubbles in the fountain region of the ionosphere. Events with quasi-regular variations in the amplitude with RMS=0.0613; 0.1579 are shown in Fig. 5.1.2 (left and right panel). The phase path excesses at two frequencies are highly correlated and changed in the interval $-0.8\ \mathrm{m} \leq \Phi_{1,2} \leq 0.6\ \mathrm{m}$. The amplitude variations in Fig. 5.1.2 (right panel) are localized between 50 and 70 km. Sharp gradients of the electron density in inclined ionospheric plasma layers may be a cause of the amplitude variations in this altitude interval where expected influence of the ionospheric D- and E-layers and neutral atmosphere is negligible. Strong ionospheric event with evident diffraction structures in RO signal is demonstrated in Fig. 5.1.3. The diffraction patterns are clearly seen at the heights 60–62.5 km. Intersection of the caustic boundary is seen at the height 60 and 62.5 km. Small-scale amplitude variations (thin features) are connected with diffraction pattern near the caustic boundary. This case can be considered as a consequence of diffraction of radio wave on sharp vertical gradients of the electron density in a sporadic E-layer E_s. Information on details of the electron density distribution can be obtained because absence of interferences connected with atmosphere above 50 km. The electron density distribution can be obtained by usual Abel's inversion with accounting for systematic changes in the height connected with horizontal displacement of the tangent point. Intensive ionospheric regular event with influence of the E_s is shown in Fig. 5.1.3 (left panel). Fine structures, which are seen in the phase path excess, are corresponding to E_s and disposed at the ray perigee heights 75 and 84 km. Usually the inclined E_s are disposed between 95 and 110 km in the morning ionosphere. The inclination of E_s can be found if the real height is known from the independent measurements. In any case the vertical distribution of the electron density and its gradients can be found by means of Abel's inversion.

Checking of CHAMP and GPS/MET data showed usefulness of the amplitude data in the height interval 40–100 km as an indicator of the current state of the ionosphere. Analysis of the amplitude data can give, in principle, the electron density distribution in sporadic E-layers and then can produce the phase and amplitude corrections at the atmospheric part of RO signal.

Analysis of CHAMP radio occultation data indicates importance of the amplitude variations for classification of the ionospheric influence on the RO signal. This classification can be mainly based on the dispersion of the amplitude variations, and on the form of spectra of the amplitude variations. Analysis revealed five types of the ionospheric influence on CHAMP radio occultation data.

(1) Noisy events with highly incoherent contribution of ionospheric disturbance to the amplitude of RO signal.
(2) Regular events with quasi-periodical nearly coherent contributions to the amplitude and phase of RO signal.
(3) Regular flashes in the amplitude and phase of RO signal.
(4) Diffractive events with clearly seen diffraction phenomena in the ionosphere.
(5) Quiet events.

Strong regular variations in the amplitude of RO signal in the most cases are connected with inclined ionospheric layers. Regular character of the main part of the ionospheric disturbances indicates a possibility to obtain additional information on the ionospheric structure from RO measurements. This reveals usefulness of RO method for global investigation of E_s in the lower ionosphere which is difficult by the Earth's based tools.

5.2. Investigation of sporadic layered plasma structures in the lower ionosphere

A description of the CHAMP RO mission is given in various publications [47]. The data archives may be accessed at: http://isdc.gfz-potsdam.de/champ. Results of CHAMP RO measurement of the phase and amplitude of the GPS signals are shown in Fig. 5.2.1 for RO event № 0001, September 21, 2003, 13 h 21 m LT with geographical coordinate 83.4° S latitude and 161.8° W longitude. The height h of the ray perigee evaluated from orbital data by use of a standard model of standard refraction in the atmosphere (2.1.4) is depicted on the horizontal axis in Fig. 5.2.1. The plotted points correspond to the sampling rate 50 Hz. The height-sampling interval is about 30 m, depending on vertical RO beam velocity. Curves 1 and 2 indicate the results of measurements of the phase path excesses at two GPS frequencies L2 and L1, respectively. For a better comparison, the phase path excess calculated by use of the refractivity model [30,31,69] has been subtracted from the phase path excesses L1, L2, and F0 (curves 1– 3 in Fig. 5.2.1, left panel). All data correspond to the altitudes retrieved by use of the Abel-inverted refractivity. The phase path excess residual F0 (curve 3) has been obtained after an ionospheric correction, i.e., after excluding the ionospheric influence by linear combination of the phase path excesses residuals at the frequencies L1 and L2. The phase residuals corresponding to neutral gas (curve 3) are changing from –1 m at 30 km to –8 m at 5 km. These values correspond to a distinction of vertical profile of refractivity in the polar atmosphere from the altitude profile of the refractivity used for simulation of the phase excess. Curve 1 in Fig. 5.2.1, right panel, demonstrates the refractive attenuation variations calculated by use of the amplitude data at the first GPS frequency F1. Curves 2–4 indicate the results of estimating the refractive attenuation by use of Eq. (4.5.1) from the phase excess data at the frequencies L2 and L1

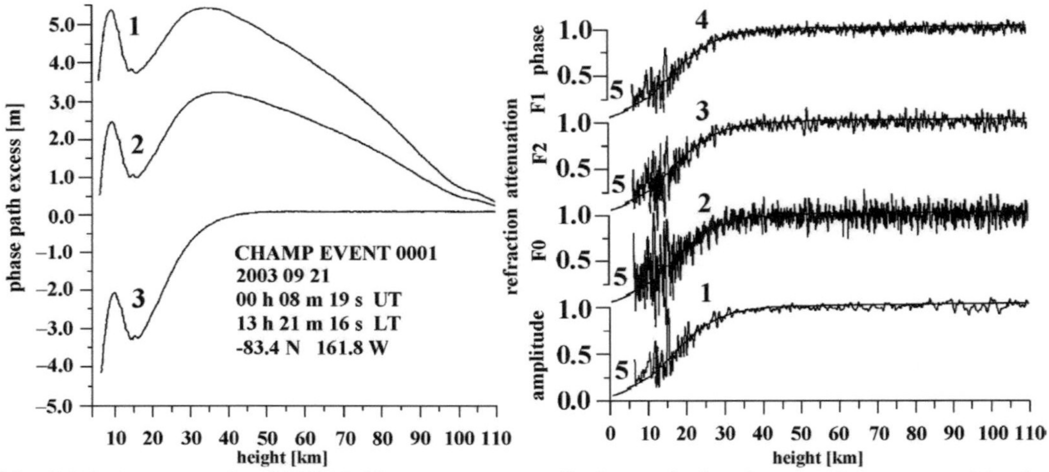

Fig. 5.2.1. An example of CHAMP measurements. Left panel: the phase path excess residuals L1 and L2 as functions of height retrieved by use of the Abel-inverted refractivity (curves 1 and 2, respectively). The phase path excess residuals L1 and L2 have been calculated by subtracting the phase excess evaluated by use of the refractivity model of the neutral atmosphere from the measured phase excesses L1 and L2. The phase path excess residual F0 (curve 3) has been obtained after ionospheric correction, i.e. after excluding the ionospheric influence by a linear combination of the phase path excesses residuals at the frequencies L1 and L2. Right panel: the refraction attenuation of the radio waves (curves 1–4) obtained from the amplitude (curve 1) and phase (curves 2–4) data as functions of height retrieved by use of the Abel-inverted refractivity. Smooth curves 5 indicate result of modeling of the refraction attenuation by use of the refractivity model of the neutral atmosphere.

(curve 3 and 4, respectively) and the phase excess data F0 obtained after the ionospheric correction. For calculation of the refractive attenuation from the phase excess data, coefficient m in Eq. (4.5.1) has been evaluated with using the satellites orbital data, which include the position and velocities of the GPS and LEO satellites relative to the center of spherical symmetry- point O (Fig. 3.4.1). Smooth curves 5 describe results of modeling of the refractive attenuation. Good correspondence between the refractive attenuations obtained from the amplitude and phase data is seen in Fig. 5.2.1, right panel (curves 1–4). Also results of modeling of the refractive attenuation coincide well with experimental data. Thus, we may conclude that the relationships (3.4.15) are valid in the RO experiments up to altitudes about 30–40 km. The relationships (3.4.15) open a new way to measure the refractive attenuation in different kinds of RO experiments including investigation of the planetary atmospheres. Thus, the eikonal acceleration has the same importance for the RO experiments as the well-known Doppler frequency. Note that by use of the phase data one may correct the amplitude data for systematic errors caused by the trends in the antenna gain and direction, and receiver's noise variations.

In the following we will consider a possibility to locate the tangent point T on the ray GTL (Fig. 3.4.1) by use of Eqs. (3.4.21)-(3.4.24). When the RO ray moves through the lower stratosphere and troposphere, the tangent point T is nearly coincides with projection Q of

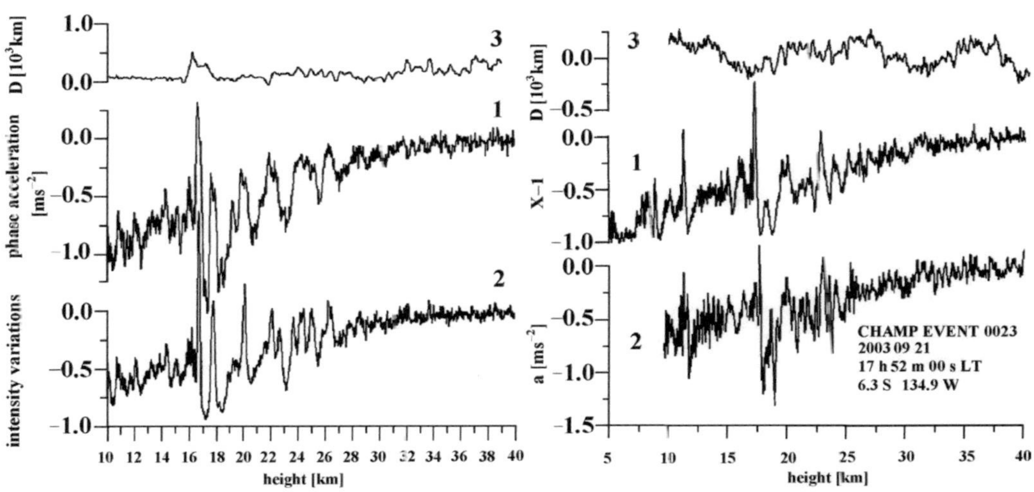

Fig. 5.2.2. Eikonal acceleration a and refractive attenuation X–1 at the first GPS frequency L1 (curves 1 and 2, respectively) for two CHAMP RO events: № 0136, January 14, 2003 (left panel) and № 0023, September 21, 2003 (right panel). Below 10 km altitude the eikonal acceleration a was not be estimated because high variability of the phase data. Curves 3 indicate the altitude dependence of distance d calculated by use of Eq. (3.4.21) with averaging on 129 samples (2.5 s time interval).

the center of the spherical symmetry- point O on the line of sight GL (Fig. 3.4.1). In the mesosphere and upper stratosphere vertical gradients corresponding to influence of neutral gas are weak and the tangent point T may be displaced because of the influence of the inclined plasma layers in the E- or F-layers of the ionosphere [156]. In this case, the center of local spherical symmetry is displaced to a point O′ and its projection to the line of sight will be disposed at a point Q′ (Fig. 3.4.1). New location of the tangent point will coincide with a point T′ at the distance d from the ray perigee T (Fig. 3.4.1). There exists a possibility to determine the distance d as a difference between the distance Q′L and QL. The distance QL may be determined from the satellites orbital data. The distance Q′L can be found from Eqs. (3.4.21)-(3.4.24) by use of known value of parameter m . To check a possibility of measuring the distance d , we will consider the CHAMP data relevant to two equatorial RO events 0136 (January 14, 2003, at 14 h 40 m local time at the geographical coordinate 3.2° S latitude and 3.8° E longitude) and 0023 (September 21, 2003, at 17 h 52 m local time at the geographical coordinate 6.3° S latitude and 139.4° W longitude), which are shown in Fig. 5.2.2 for the 10–40 km (left panel) and 5–40 km (right panel) height interval, respectively. The eikonal acceleration a_{Φ} calculated as the second derivative of the phase-path excess on time and intensity variations $1-X$ at the frequency L1 are shown in Fig. 5.2.2 (curves 1 and 2, respectively). As seen in Fig. 5.2.2, there is a good correspondence between variations of the eikonal acceleration and intensity of RO signal. Coefficient m is different in RO events № 0136 and № 0023. Average ratio of the refractive attenuation and the eikonal acceleration m is about 1.0 [s^2/m] in the 5–40 km height interval for the RO event 0136. This value m is about 1.5 times greater than that for RO event 0136. Curve 3 in Fig. 5.2.2 indicates the deflection D of point T from its standard location in the atmosphere. Deflections in the

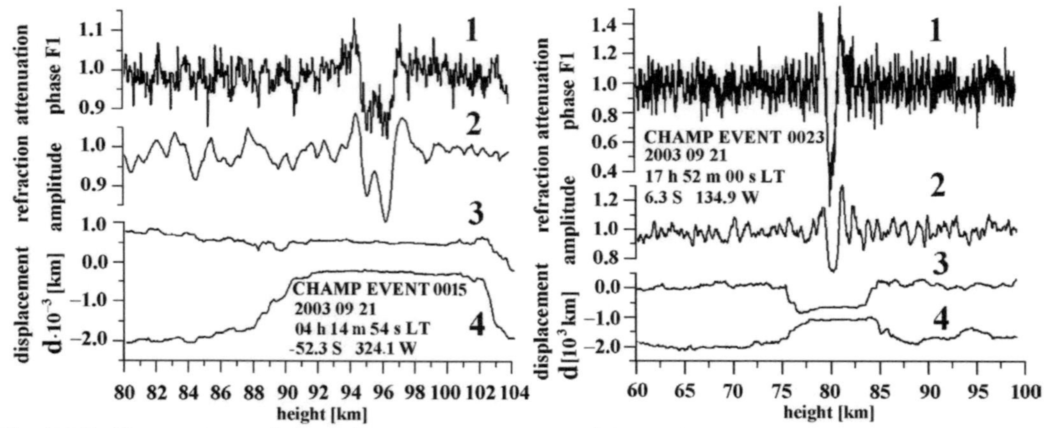

Fig. 5.2.3. Comparison of the refractive attenuation of the radio waves found from the phase and amplitude data (curves 1 and 2) with results of the distance d estimation by use of Eqs. (3.4.23) (curve 3) and (3.4.24) (curve 4) for CHAMP RO event 0015, September 21, 2003 (coordinate 52.3 S, 324.1 W, 04 h 14 m 54 s LT) (left panel) and 0023, September 21, 2003 (coordinate 6.3 S, 134.9 W, 17 h 52 m 00 s LT) (right panel).

direction to point G or to point L are positive or negative, respectively. Deflection of point T between 10 and 20 km from its standard position is mostly small ~±150 km. This indicates the atmospheric origin of the phase and amplitude variations of RO signal at these altitudes. A correlation between the refractive attenuation and eikonal acceleration is clearly seen between 10 and 30 km as shown above. Slow regular changes of the refractive attenuation and eikonal acceleration introduce main contribution to this correlation. Below 10 km , this correlation is destroyed by multipath propagation in the troposphere. Above 30 km the ionospheric effect becomes important. Below we will consider a possibility to use the ionospheric effect for locating the ionospheric layers. Examples of the refractive attenuation recalculated from the amplitude and phase data X_a and X_p are shown in Fig. 5.2.3 by curves 1 and 2. The forms of the refractive attenuations X_a and X_p are similar and indicated ionospheric layers at the altitudes of rayperigee 95 km and 80 km (left and right panels, respectively). The width of both ionospheric layers, as estimated from curves 1 and 2, is about 2 km. Altitude 80 km is unusual for ionospheric layers. Thus, in this case the center of spherical symmetry is displaced and the altitude of the ionospheric layers should be recalculated. Results of estimation of the upper and lower boundaries for displacement D with averaging on 2.5 s time interval (M= 65 samples) are indicated by curves 3 and 4, correspondingly, for CHAMP RO events 0015 and 0023. For event 0015, the upper and lower boundaries for displacement d near altitude 95 km are about 450 km and –500 km (curves 3 and 4, respectively, Fig. 5.2.3, left panel).

One can estimate the displacement d and parameter m by use of Eqs. (3.4.21), (3.4.23) and (3.4.24) near the center of layer at altitude h equal to 95 km (Fig. 5.2.3, left panel). Results of determination of parameter m , displacement d and correction to the layer altitude Δh are given in the Table 5.2.1 as function of the altitude h of ray perigee T. Also the refractive attenuation X_a , estimated from the amplitude data, and refractive attenuation X_p

calculated from Eq. (4.5.1) by use of the measured eikonal acceleration a and estimated value of parameter m are given in the Table 5.2.1. The refractive attenuations X_a and X_p are nearly equal, thus indicating the small magnitude of displacement d . The displacement d as seen from the second column of the Table 5.2.1 is changing in the –180 km– +65 km interval and corresponding corrections to the altitude Δh are about 0.3 km in average. Therefore, the results of parameter m estimations by use of equation (3.4.24) confirm location of sporadic E-layer at the height 95.5 km with small displacement d ~ –30 km from ray perigee T (Fig. 3.4.1). One can estimate by use of data in Table 5.2.1 the influence of random errors on the evaluation of distance d from Eqs. (3.4.21) and (3.4.22). Parameter m changes in the 0.3349– 0.3780 [s²/m] interval with statistical error ±0.028 [s²/m] and relative error ±8.1%. The estimated distance d changes in the -179.1 – +64.9 km interval with statistical

Height h [km]	Distance d [km]	$\Delta h = \frac{d^2}{2r}$ [km]	Refractive attenuation (F1 phase data) X_p	Refractive attenuation (F1 amplitude data) X_a	m [s²/m]
95.746	52.9	.21	.8987	.8956	.3762
95.700	9.3	.00	.9010	.9000	.3685
95.655	-26.5	.05	.9031	.9038	.3621
95.610	61.5	.29	.9096	.9065	.3776
95.564	8.6	.00	.9094	.9086	.3683
95.519	-121.7	1.15	.9051	.9102	.3452
95.474	43.5	.14	.9145	.9122	.3744
95.428	28.9	.06	.9143	.9127	.3718
95.383	-121.5	1.15	.9098	.9146	.3452
95.338	-14.5	.01	.9152	.9154	.3641
95.292	47.0	.17	.9201	.9179	.3749
95.247	-105.4	.86	.9179	.9216	.3480
95.201	-59.2	.27	.9239	.9256	.3561
95.156	64.9	.32	.9323	.9298	.3780
95.111	-179.1	2.50	.9317	.9373	.3349
95.065	-33.8	.08	.9428	.9434	.3605

Table 5.2.1. Displacement d and correction Δh in the height of tangent point

error ±65 km. The accuracy in the evaluation of distance d by considered technique corresponds to the theoretical limit of horizontal resolution of RO method estimated previously as ±100 km. It follows that the estimation of parameter m from Eq. (3.4.22) is more precise than from Eqs. (3.4.23), and (3.4.24) in the case of strong scintillations of the amplitude and phase caused by layered structures in the ionosphere.

Altitude 80 km observed in the event 0023 (Fig. 5.2.1, right panel, curves 1 and 2) is unusual for a sporadic ionospheric layer. Thus, in this case, the center of spherical symmetry is

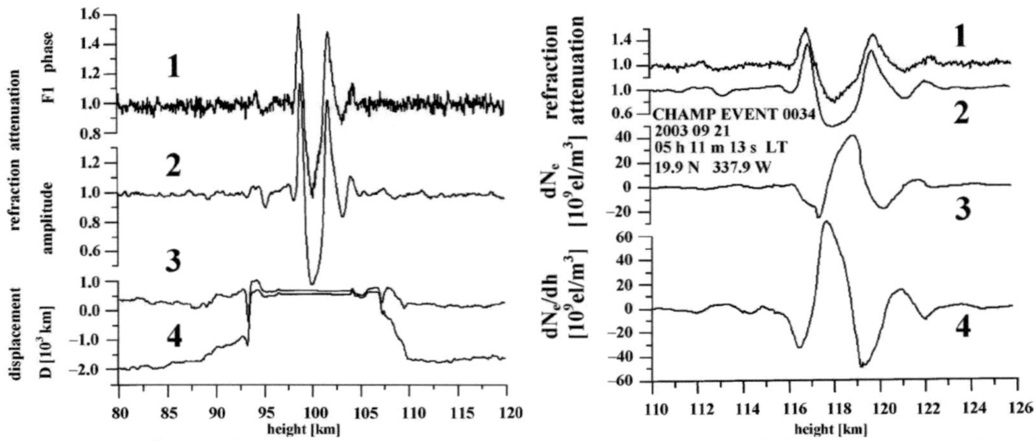

Fig. 5.2.4. Left panel: Comparison of the refractive attenuation of the radio waves found from the phase and amplitude data (curves 1 and 2) with results of distance d estimation by use of Eqs. (3.4.23) (curve 3) and (3.4.24) (curve 4) for the CHAMP RO event 0034, September 21, 2003 (coordinate 19.9 N 227.9 W, 05 h 11 m 13 s LT) as function of the altitude of the ray perigee T. The time interval for averaging was equal to 2.5 s. Right panel: Comparison of the refractive attenuation of the radio wave found from the phase and amplitude data (curves 1 and 2) with results of estimation of the electron density and its vertical gradient variations (curves 3 and 4) as function of the real altitude of the sporadic E-layer $h+\Delta h$.

displaced and the altitude of ionospheric layer must be recalculated. The large difference (about two times) between the refractive attenuation X_a and X_p calculated from the amplitude and phase data under the assumption of the spherical symmetry with the center at point O (Fig. 3.4.1) (curves 1 and 2 in Fig. 5.2.3, right panel) supports this conclusion. Curves 3 and 4 in Fig. 5.2.3 (right panel) indicate the displacement d in the interval from –600 km (upper boundary) to –900 km (lower boundary). A more careful investigation by use of Eq. (3.4.24) gives value d in the –680 km - –630 km interval, and correction to altitude Δh is about 30 km. Thus, the real height of E_s is about 110 km. The layer is displaced by 650 km to receiver from the ray perigee T and has inclination $\delta \approx 6°$ to the local horizontal direction. The height of sporadic E-layer is increased in the receiver direction. The lower boundary of the horizontal length, L_e, of sporadic E-layer can be estimated from relationship $L_e \approx \Delta d / \delta$, where Δd is the layer width estimated on dependence of the refractive attenuation on height. For event 0023, value $\Delta d \approx 3.5$ km and the horizontal length of E-layer is greater than 35 km.

Another example of intensive E_s is given in Fig. 5.2.4. Curves 1 and 2 (Fig. 5.2.4, left panel) indicate the refractive attenuations X_a and X_p, respectively, as the functions of altitude h of ray perigee T (Fig. 3.4.1). Curves 3 and 4 demonstrate the upper and lower boundaries, respectively, for displacement d (Fig. 5.2.4, left panel). As seen in Fig. 5.2.4 (left panel), both minima in the refractive attenuation found from the amplitude and phase data correspond to height $h \approx 100$ km. The upper and lower boundaries of the displacement d are nearly

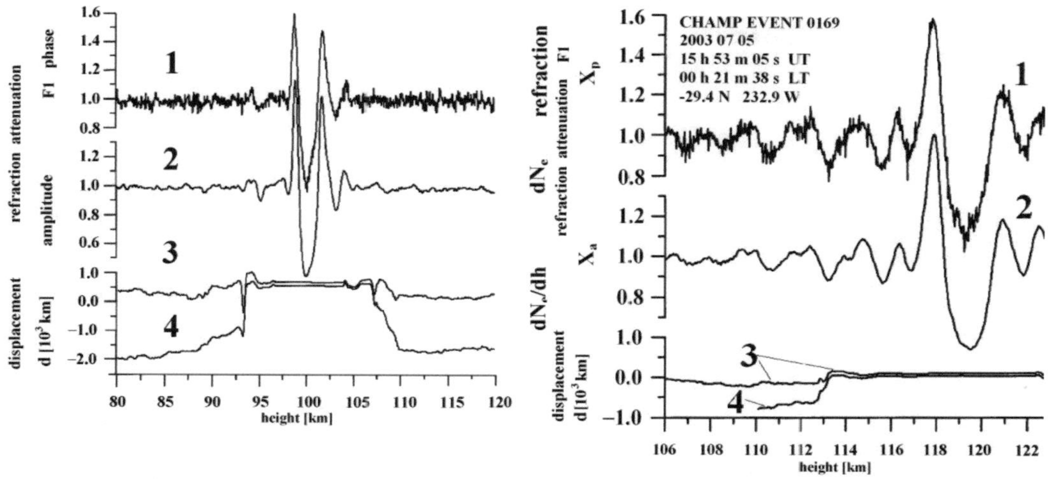

Fig. 5.2.5. Left panel. Comparison of the refractive attenuation X_a and X_p retrieved from the amplitude and phase variations at the GPS frequency L1 (curve 1 and 2), and results of determination of the displacement of point T from RO ray perigee (curves 3 and 4) for CHAMP RO event 0169, July 05, 2003, 29.4° S 232.9° W. Right panel. The refractive attenuation X_a and X_p retrieved from the amplitude and phase variations (curve 1 and 2), and retrieved variations of the electron density and its vertical gradient as functions of the corrected height in the ionosphere for CHAMP RO event 0169.

coinciding and equal to 500 km (curves 3 and 4, Fig. 5.2.4, left panel). This value d corresponds to location of the tangent point T′ in sporadic E-layer between transmitter G and ray perigee T at altitude $h + \Delta h$, where $\Delta h \approx 18$ km. By use of theoretical relationships published in [55], it is possible to retrieve variations of the electron density $dN_e(h)$ and its gradient $dN_e(h)/dh$ from variations of the refractive attenuation, indicated by curves 1 and 2 in Fig. 5.2.4. Restored variations of the electron density $dN_e(h)$ and its gradient $dN_e(h)/dh$ are shown in Fig. 5.2.4, right panel, by curves 3 and 4, respectively. The electron density variations are concentrated at 116–120 km in the interval ±30 $[10^9 m^{-3}]$ and vertical gradient of the electron density changes in the ±60 $[10^9 m^{-3} km^{-1}]$ interval . Maximum in the electron density variation is about of 40 $[10^9 m^{-3}]$ and is displaced by 1.5 km relative to minimum of the refractive attenuation disposed at 118 km altitude . Maximum in vertical gradient of the electron density is about 70 $[10^9\ m^{-3} km^{-1}]$ and locates by 0.5 km below relative to minimum of the refractive attenuation. Additional example of application of the eikonal acceleration/intensity ratio technique to analysis of the amplitude and phase variations is given in Fig. 5.2.5. Sporadic E- layers contributions to variations of RO signal are considered in detail in Fig. 5.2.5 (left panel). The displacement can be estimated by using curves 3 and 4. The centre of sporadic E-layer is located at 119.5 km altitude and practically coincides with the RO ray perigee since the distance d (curve 3 and 4 in Fig. 5.2.5, left panel) is equal to 0. The accuracy in determination of value d , which may be estimated as a difference between curves 3 and 4 in Fig. 5.2.5, is about 30–100 km. Results of restoration of the electron density variations $dN_e(h)$ and its gradient $dN_e(h)/dh$ from the amplitude variations of RO signal are given for the considered RO event 0169 in Fig. 5.2.5 (right panel). The initial data: the eikonal acceleration a_Φ and refractive attenuation $1-X$, are shown by curves 1 and 2. Curves 3 and 4 demonstrate the retrieved variations $dN_e(h)$ and $dN_e(h)/dh$ as functions of the corrected height h. Both curves 1 and 2 are nearly coinciding. This is a new validation of the Eq. (4.5.1) for the case of sporadic E-layers in the ionosphere. The electron density perturbations $dN_e(h)$ changes mainly in the $\pm 5 \cdot 10^9$ – $\pm 25 \cdot 10^9$ [m^{-3}] interval, and its vertical gradient $dN_e(h)/dh$ in the $\pm 20 \cdot 10^9$– $\pm 50 \cdot 10^9$ [$m^{-3}km^{-1}$] interval. The estimated heights and plasma parameters are usual values for the sporadic E-layer in the ionosphere [157]. Therefore, the suggested method has a promise to be an effective tool for the localization of the sporadic E-layers and inclined plasma layers in the ionosphere.

5.3. Comparison with the ground-based observation data

The CHAMP RO amplitude variations for some RO events, which can be recognized as C- and S- types of the amplitude scintillations observed earlier in the communication link Inmarsat - ground station at frequency 1.5 GHz [158], are shown in Fig. 5.3.1. The noisy C-type amplitude variations in the RO signals are given in Fig. 5.3.1, top panel. These scintillations have been observed in the equatorial region at local evening during RO events No. 0051, November 19 (curve 1); No. 0053, July 05 (curve 2); No. 0069, February 24 (curve 3); No. 0135, July 04 (curve 6), 2003 year. Average values of the S_4 index are equal to 18.8%, 32.0%, 20.2% and 50.5% (curves 1,2,3 and 6, respectively). Event 0159 corresponds to local morning (May 03), in the polar ionosphere with S_4=16.0% (curve 4). Curve 4 corresponds to the daytime noisy event near the north geomagnetic pole. Curve 5 relates to quiet nighttime ionosphere with S_4=3.0% (event No. 0198, May 03), when the

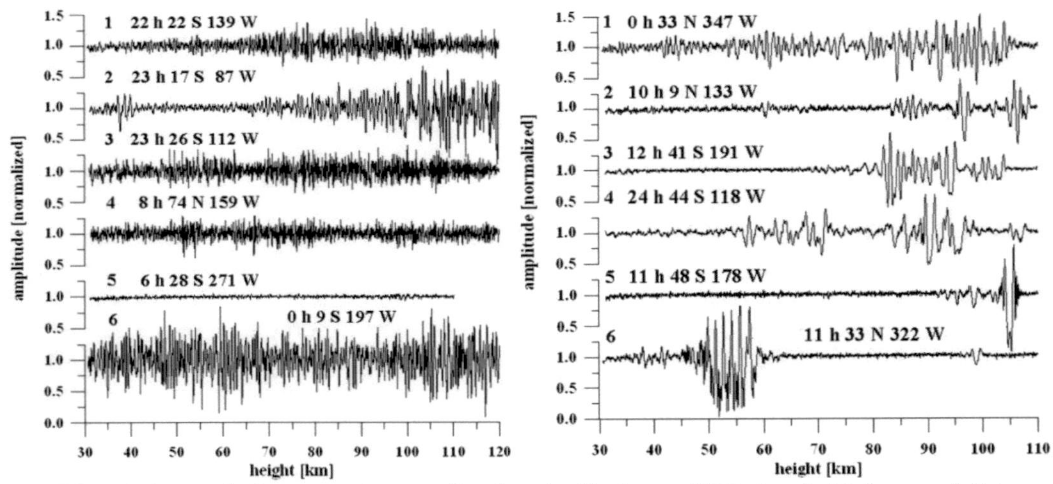

Fig. 5.3.1. Left panel: C-type noisy amplitude scintillations of RO signal. Right panel: S-type quasi-regular amplitude variation. Legends indicate the local time (LT) and the geographical coordinates of RO measurements

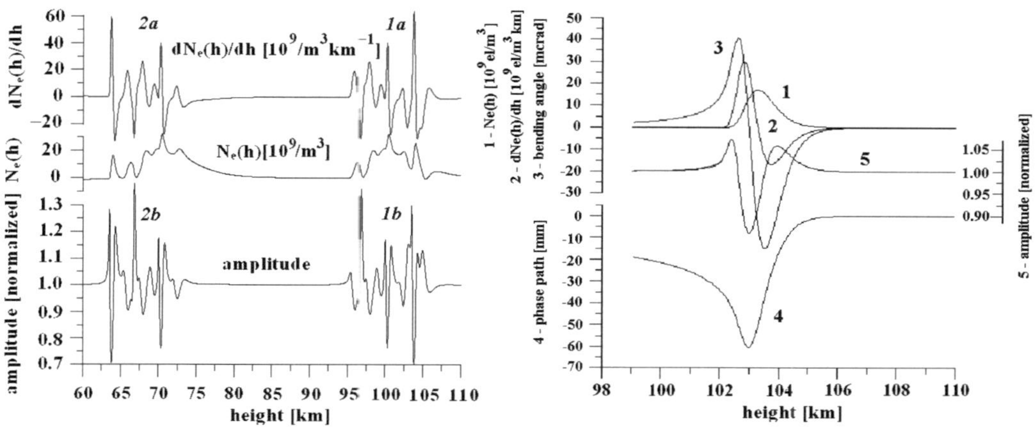

Fig. 5.3.2. Results of modeling the amplitude of RO signal and the model of the electron density $dN_e(h)$. Left panel . Model consist of seven independent layers $dN_e(h)$. Right panel . Model consists of one layer $dN_e(h)$.

amplitude fluctuations in the height interval from 30 to 110 km were caused mainly by random noise of the receiver. The geographical position and local time of the noisy RO events correspond to the same parameters of the noisy amplitude scintillations observed earlier in satellite-earth links [158]. Results of simulation of the local mechanism of ionospheric influence on the amplitude of RO signal are shown in Fig. 5.3.2. Curves 1a, 1b describe model of the electron density distribution in two patches of sporadic E-layers with horizontal inclination 0° and 5°. Both sporadic E-layers are disposed at the same height 96–104 km, but the second patch is displaced along the ray path by 500 km relative to the first one.The electron density distribution for the first patch (curve (1a) in Fig. 5.3.2, left panel) is displaced for better comparison with the second patch (curve 2a). The influence of the first patch on the amplitude variations (curve 1b) is seen in the height interval 96–106 km, the influence of the second patch (curve 2b) is seen in the height interval 63– 75 km. As seen in Fig. 5.3.2, left panel, the width of vertical distribution of the amplitude variations coincides approximately with the size of vertical interval of the $dN_e(h)$ variations. The altitude profiles of the electron density (curve 1), its vertical gradient (curve 2), and corresponding height dependences of the bending angle (curve 3), phase path excess (curve 4), and normalized amplitude (curve 5), evaluated for single ionospheric layer is shown in Fig. 5.3.2 (right panel). The width of vertical distribution of the amplitude variations is nearly equal to vertical size of the layer measured as the width of the altitude profile of $dN_e(h)/dh$. The form of the amplitude reply is practically the same as S- type variations, observed in [158] by use of the ground based tools. It follows that the proposed local mechanism of ionospheric influence on the RO signal can be used also for analysis of the amplitude scintillation observed in satellite-earth links. The noisy scintillations can be associated with small-scale plasma irregularities in the F- or E- layers of the ionosphere [158].

5.4. Variations of the S_4 index and solar activity

As stated above the amplitude scintillations are very important because in the case of layered structures they are directly connected from one side- with the eikonal acceleration and from another side- with Vertical gradients of refractivity in the plasma layers. The emphasis in the RO studies is now concentrating on the phase variations of RO signals in the trans-ionospheric links because direct connection with total electron content [156, 159–161] and on analysis of the amplitude variations owing to their dependence on vertical gradient of refractivity [65, 66, 73, 84, 85, 123, 124, 135, 162–165]. It follows that (1) the S_4 index of amplitude variations can be considered as an index of the ionospheric plasma influence on RO signal in the trans-ionospheric satellite-to-satellite links in a like fashion with the S_4 index introduced formerly for the trans-ionospheric satellite-to-Earth links [167]; (2) the S_4 index can be used in the satellite-to-satellite links as a radio-physical index of activity of plasma disturbances in the ionosphere; and (3) the relative number of GPS RO events with high values of the S_4 index in the satellite-to-satellite links can be used to establish a connection between the intensity of plasma disturbances and solar activity.

We will describe the amplitude variations of RO signal by the magnitude of the S_4 scintillation index

$$S_4 = \left[<(I - <I>)^2 \right]^{1/2} / <I>, \tag{5.4.1}$$

where < > is the average relevant to the height of the radio ray perigee above 40 km, and $I(t)$ is the intensity of RO signal. When the amplitude and phase variations are caused by layered ionospheric structures, the S_4 scintillation index can be connected with the refractive attenuation X–1 and the eikonal acceleration variations a (equation (6)):

$$S_4 = \left[<(X-1)^2> \right]^{1/2} = m \left[<a^2> \right]^{1/2}. \tag{5.4.2}$$

The relationships (3.4.15) and (5.4.2) give new possibility to reveal the origin of the phase and amplitude scintillations in the near-Earth propagation medium and may have a general significance for the trans-ionospheric satellite-to-satellite and satellite-to-Earth links. In the following we will indicate that the S_4 scintillation index is a key parameter connecting the amplitude variations of RO signal with space weather on a global scale.

Influence of sporadic solar activity can be considered in the case of solar flare during October 29- 31, 2003. To analyze in more details the geographic and time distributions of the amplitude scintillations in RO signal, we treat about 2000 CHAMP RO events from October 27 to November 9, 2003. The histograms of the integral distribution of the S_4 index, which have been obtained for each day of measurements, are shown in Fig. 5.4.1. The ratio of the number of cases with the magnitude S_4 larger, than value plotted on the abscissa, to the total number of measurements for the corresponding day, is plotted on the ordinates in Fig. 5.4.1. The histograms in the right panel of Fig. 5.4.1 correspond to data of S_4 measurements on October 29–31, as well as on October 28, November 1, and November 4, 2003, performed during relatively intense and weak disturbances in RO signal, respectively. The initial histogram values for the entire period of observations are coincident and correspond to an magnitude S_4 of ~2.8%, which depends on the level of fluctuations of receiver noise. Three days, October 29-31, with considerably higher S_4 levels than on the remaining days are distinguished in Fig. 5.4.1. The histograms for this time interval are much higher than for the remaining days.

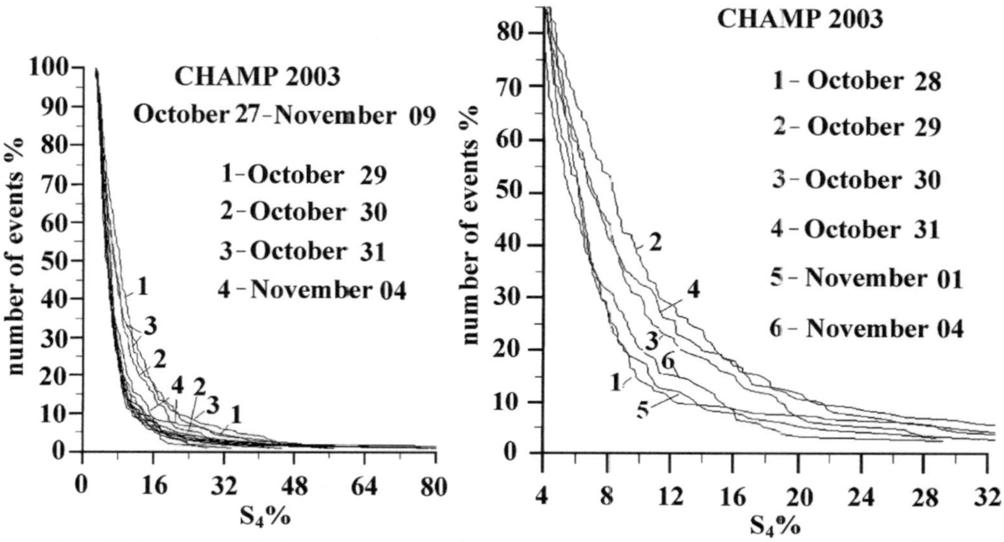

Fig. 5.4.1. Histograms of the S_4 index variations corresponding to perigee altitudes of the CHAMP RO signal trajectory (point T in Fig. 3.4.1) higher than 40 km. The curves corresponding to the histograms for October 27 and 28 and for November 1-3 and 5-9, 2003, are almost indiscernible in the left-hand panel. The right panel demonstrates the histograms in more details.

The most difference between the histograms is observed in the 5%-19% interval of the S_4 index with maximum about of 12% (Fig. 5.4.1, left and right panels). Value 12% of the S_4 index is by a factor 4.3 greater than the level of the receiver's noise. Therefore, in the CHAMP RO experiments at the first GPS frequency values of S_4 index in the 5%-19% interval can be considered as the most sensitive for estimation of the ionospheric influence on RO signal. An analysis performed makes it possible to introduce RO index of ionospheric activity R_i. By definition, the R_i index is equal to the ratio of the number of RO events with the S_4 value of RO signal at $h > 40$ km larger than three- four magnitudes of S_4 caused by noise of the transmitter- receiver system to the total number of RO events for a certain time interval (e.g., for a day). The R_i variations for the period from October 27 to November 9, 2003 are shown in left panel of Fig. 5.4.2. The histograms in Fig. 5.4.1 and Fig. 5.4.2 are in good agreement with variations in the hourly index of the solar wind magnetic field (*DST*, nT) obtained from data of the satellite measurements. During October 26-28 and November 1-3 and 5-9, the *DST* index was almost constant, which corresponds to decreased values of the integrated distributions in the histograms (Fig. 5.4.1 and Fig. 5.4.2 (left panel). Abrupt changes in the index of the solar wind magnetic field were observed on October 29-31 and

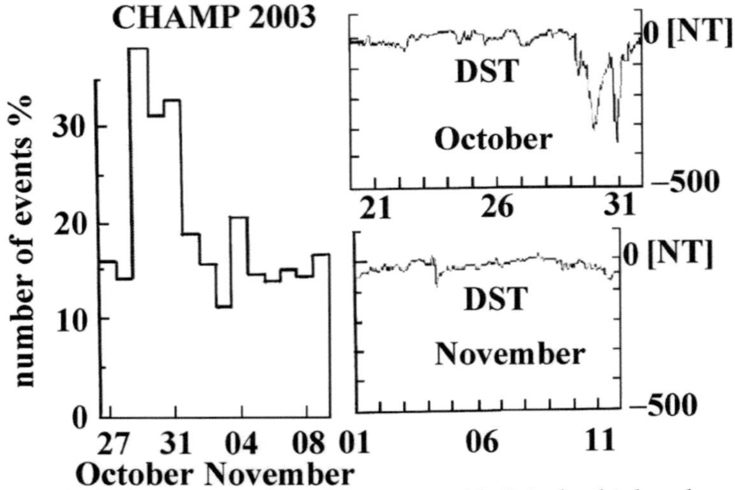

Fig. 5.4.2. Left panel: the histogram of the RO events with S_4 index higher than 5% according to data of the CHAMP measurements performed from October 27 to November 9, 2003. Right panel: the dynamics of the hourly index of the solar wind (*DST*, nT) according to data from http://swdcwww.kugi.kyoto-u.ac.jp/aeasy/index.html for October 26–31 and November 1–11, 2003.

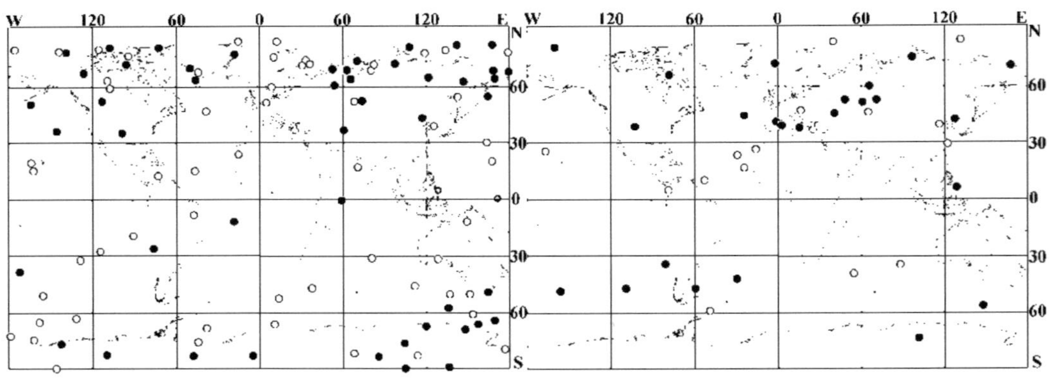

Fig. 5.4.3. The geographic distribution of the RO events with S_4 index value higher than 12% on October 26–28, 2003 (right panel) and October 29–31, 2003 (left panel). Open and filled circles correspond to the nighttime (from 2000 to 0800 LT) and daytime (from 0800 to 2000 LT) local time intervals, respectively.

November 4, which corresponds to a twofold increase in the number of events with the increased (larger than 12%) S_4 index of variations in RO signal.

The geographic distributions of increased (greater than 12%) S_4 index of variations in RO signal amplitude on October 26–28 and on October 29–31, 2003 are compared in Fig. 5.4.3.

According to Fig. 5.4.3 (left panel), the general number of events with S_4 index values greater than 12% increased by 2.5 times in period October 29–31 as compared with period October 26–28. The most important changes occurred mainly in the polar regions in Antarctica and in the Northern Hemisphere near geomagnetic South and North poles. It is interesting to note that ionospheric disturbances are concentrated above continents near coasts and in mountain regions (eastern coast of Asia, Japan, Kamchatka, western regions of South America, etc.). This corresponds to observations made previously in [160].

5.5. Geographical and seasonal distributions of CHAMP RO events with intense amplitude variations

The geographical distributions of the intense ionospheric events (with the S_4 index greater than 0.12) for all types of the amplitude scintillations in the CHAMP RO signals at 1575.42 MHz are demonstrated in Fig. 5.5.1 and Fig. 5.5.2. Note, that these distributions may be linked to changes in sampling by CHAMP. Our analysis showed that the satellite coverage over the relevant periods was nearly regularly distributed. Therefore these distributions

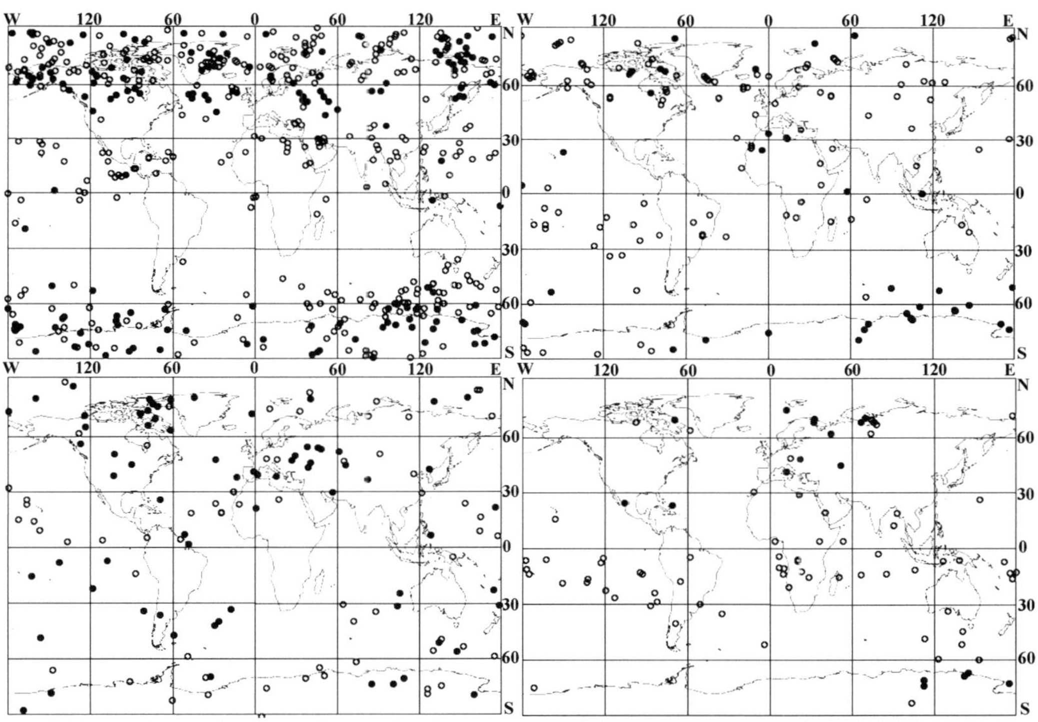

Fig. 5.5.1 . Maps of intense ionospheric events with S_4 index greater than 0.12 for October 2001 (left top panel), October 2002 (right top panel), October 2003 (left bottom panel), and April 2004 (right bottom panel). The circles show the geographical position of the tangent point *T* (Fig. 3.4.1). The day-time events are marked by closed circles (local time 08–20 hours) and the night-time events are indicated by open circles (local time 20–08 hours).

contain important information on the spatial dependence of the strong ionospheric events during the considered time intervals. The distributions of the ionospheric events indicate that they are concentrated in some regions (e.g., the equatorial and geomagnetic North and South polar zones in Fig. 5.5.1 and Fig. 5.5.2). High activity in some equatorial regions may be connected with the evening ionospheric disturbances that arise after sunset, 20–24 hours local time, in accordance with the ground based measurements reviewed earlier [167].

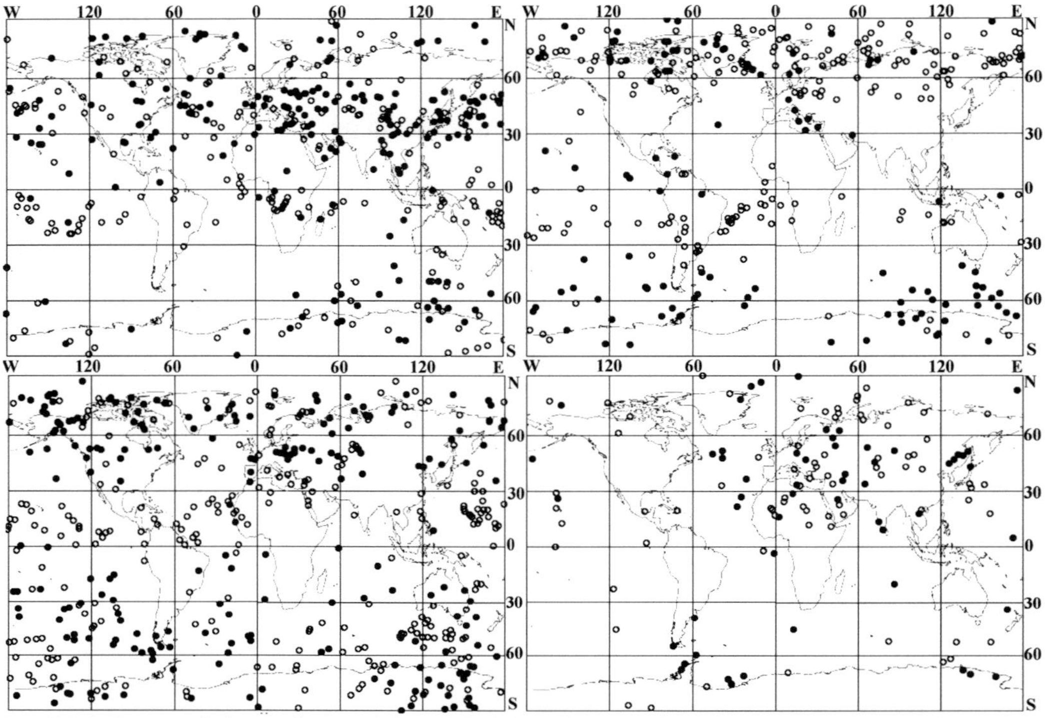

Fig. 5.5.2. Seasonal dependence of the global distribution of CHAMP RO events with intense amplitude variations (with magnitude of S_4 index greater than 0.2) for periods May 14–July 14, 2001, (left top panel), November-December 2001 (right top panel), October 28-November 26, 2003 (left bottom panel), July 2004 (right bottom panel). The day time events are marked by closed circles (local time 08–20 hours), the night-time events are indicated by open circles (local time 20–08 hours).

Note that the geographical distributions of the night- and day-time events are very different. The day-time events prevail in the North and South Polar zones. The night-time events are concentrated mainly in the equatorial and moderate latitude areas. The most significant difference in the geographical distributions of night-time (open circles) and day-time (closed circles) events is observed in April 2004 (Fig. 5.5.2, right bottom panel), and in October 2002 (Fig. 5.5.1, right top panel). This indicates different origin of the day-time and night-time plasma disturbances in the ionosphere. The number and geographical distributions of intense ionospheric events depend on time. The maximal number of events can be seen in May–July, October–December 2001 (Fig. 5.5.1, left top panel, Fig. 5.5.2, left and right top panels, respectively). The minimal number of events corresponds to October 2002 (right top panel in Fig. 5.5.1), and April and July 2004 (bottom right panels in Fig. 5.5.1 and Fig. 5.5.2, respectively). As seen in Fig. 5.5.1 and Fig. 5.5.2, the number of intense ionospheric events and their intensity decreases as time from May, July, and October 2001 to April–July 2004. This is possibly connected with a gradual diminishing in solar activity from maximal values in 2001 to minimal values in 2004. One can estimate the seasonal dependence in the geographical distributions of intense ionospheric events registered by the CHAMP satellite (Fig. 5.5.2). The seasonal displacement of the regions with intense ionospheric events in the south and north directions during the periods from May–July 2001 (Fig. 5.5.2, left top panel) to November–December 2001 (Fig. 5.5.2, right top panel), and to September 2001 (Fig. 5.5.1, left top panel) can be noted. The number of intense ionospheric events increases in the North Polar Region with latitudes greater than 60° N as the time changes from May–July 2001 to November–December 2001. Also the number of intense ionospheric events decreases at the moderate latitudes in the 30°– 60° N interval for the same time interval. The possible cause of the seasonal dependence may be connected with two important mechanisms governing the ionospheric disturbances. The first one is the ionizations caused by energetic electrons in the polar regions; the second is due to solar radiation. The influence of solar radiation has a seasonal character because ionization in the ionosphere follows annual and diurnal motion of the ionospheric sub-solar point. As follows from this consideration, the amplitude part of the RO radio holograms is appropriate for studying the geographical and seasonal distributions of the ionospheric disturbances with global coverage and establishing their connections with solar activity.

The RO signal variations are an indicator of plasma disturbances in the ionosphere. The correlated eikonal acceleration and intensity variations of RO signal are mainly affected by inclined ionopheric layers where an large electron density gradient is perpendicular to a radio wave propagation trajectory. The eikonal acceleration/intensity ratio technique is a new method for locating the inclined layered structures (including sporadic E_s layers) in the ionosphere. This technique gives the quantitative connection between the phase and amplitude variations and allows establishing the origin of scintillation of radio signals in the trans-ionospheric satellite-to-satellite and satellite-to-Earth links.

An analysis of the histograms of RO signal amplitude variations corresponding to the solar event October 29–31, 2003 indicates a possibility to introduce the RO index of ionospheric activity R_i, which is in good agreement with changes in the hourly index of DST related to the solar wind. The analysis indicates the importance of RO studies for revealing the interrelations between the solar activity and natural processes in the ionosphere and mesosphere.

6. Expending Applicable Domain Of RO Technique

6.1. Reflections from the Earth surface

Recovering the reflected signals in MIR/GEO (wavelength 2 and 32 cm), GPS/MET and CHAMP (wavelength 19 and 24 cm) radio occultation experiments opened new perspectives for bistatic monitoring of the Earth at small elevation angles. Experimental observation of the propagation effects at low elevation angles has also principal importance for fundamental theory of radio wave propagation. During existing and future RO missions progress of bistatic monitoring of the earth's surface at grazing angles is expected. Aim of this chapter consists in analysis of radio waves propagation effects measured in experiments, conducted during MIR/GEO and GPS/MET missions at wavelength 2, 32, 19 and 24 cm.

Important connections between the Doppler shift and refractive attenuations of the direct and reflected signals are revealed. These connections allow recalculating the Doppler frequency to the refractive attenuation and open a possibility to measure the total absorption of radio waves in the atmosphere.

6.1.1. Features of direct and reflected radio wave propagation

Earlier two-position earth's radio-location experiments [30, 31] were essentially based on a technique developed previously for exploration of the Moon and Venus [12, 17]. When applied to the earth, this technique has to be refined and modified to take into account both the significantly wider variation in the earth's natural cover and the specific way that radio wave propagate in the terrestrial atmosphere at grazing angles [72, 74]. These conditions differ drastically from those found in the Venusian atmosphere, which always exhibits the critical level of refraction, and where the refraction effects show up at their strongest. In the terrestrial atmosphere, the critical refraction level does not exist under average conditions, so one can trace the reflected signal up to the radio shadow zone and observe its interference with the direct signal. The aim of this section is to analyze the conditions in which the reflected and direct signals propagate near the radio shadow zone.

6.1.2. Phase path, refractive angle and refractive attenuation

The geometry of radio waves propagation is shown in Fig. 6.1.1. Radio waves are emitting by a satellite located at point G and then are propagating along two paths- 1 and 2. Reflected and direct radio waves were registered by a receiver installed on a Low Earth Orbital (LEO) satellite located at point L. The transmitter and receiver were disposed at the distances R_C, R_L from the center of spherical symmetry (point O). The angle between the directions OG and OL is equal to θ. The transmitter and receiver have different velocities v_C, v_L. Point D is the specular reflection point disposed on a smooth sphere of radius a corresponding to the average earth's surface in the RO region.

The direct and reflected signals are propagating along the different parts through the ionosphere and atmosphere and the central angle θ is common for both trajectories GL and GDL. In the case of spherical symmetry θ is connected with the refractive angles $\xi_d(p_d)$, $\xi_r(p_r)$ and impact parameters p_d, p_r corresponding to the direct and reflected signals via Eqs.

$$\theta = \pi + \xi_d - \sin^{-1}[p_d / n(R_L)R_L] - \sin^{-1}[p_d / n(R_G)R_G], \qquad (6.1.1)$$

$$\theta = \pi - 2(\psi - \xi_r) - \sin^{-1}[p_r / n(R_L)R_L] - \sin^{-1}[p_r / n(R_G)R_G], \qquad (6.1.2)$$

$$\theta = \pi - \sin^{-1}(p_s / R_1) - \sin^{-1}(p_s / R_2), \qquad (6.1.3)$$

$$\psi(p_r) = \cos^{-1}[p_r / n(a)a], \qquad (6.1.4)$$

where $\varphi(p_r)$ is the grazing angle corresponding to the reflected ray at the point D, p_s is the impact parameter corresponding to the line of sight GL, $n(R_G)$, $n(R_L)$, $n(a)$ are the refractive indexes at point G, L, and D, respectively.

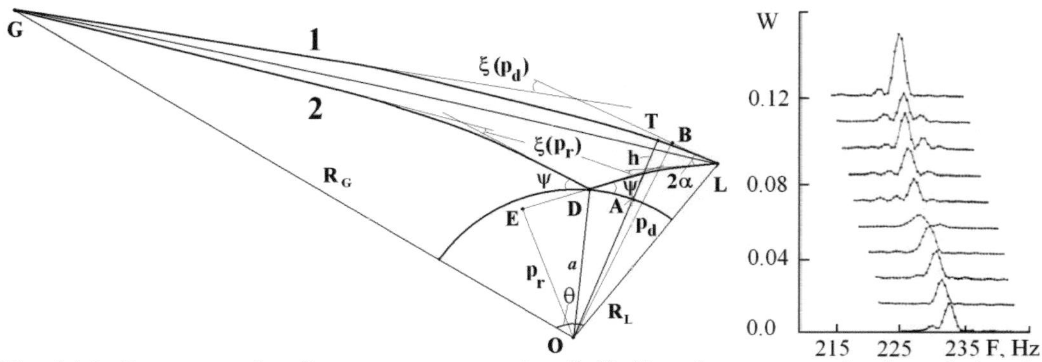

Fig. 6.1.1. Geometry of radio waves propagation (left). Doppler spectra of reflected signals (right).

The phase path excesses of the reflected and direct signals $S_{r,d}$ may be described by formulas:

$$S_r=\sqrt{n^2(R_L)R_L^2-p_r^2}+\sqrt{n^2(R_G)R_G^2-p_r^2}-2\sqrt{n^2(a)a^2-p_r^2}+p_r\xi_r(p_r)+\kappa_r(p_r)-S_s(p_s)\,, \quad (6.1.5)$$

$$S_d=\sqrt{n^2(R_L)R_L^2-p_d^2}+\sqrt{n^2(R_G)R_G^2-p_d^2}+p_d\xi_d(p_d)+\kappa_d(p_d)-S_s(p_s)\,, \quad (6.1.6)$$

$$S_s(p_s)=\sqrt{R_L^2-p_s^2}+\sqrt{R_G^2-p_s^2}\,, \quad (6.1.7)$$

where $\kappa_r(p_r)$, $\kappa_d(p_d)$ are the main refractive part of the phase path excess connected with influence of vertical gradients of refractivity in the atmosphere along the ray paths of the direct and reflected rays, $S_s(p_s)$ is the phase path in free space along the straight line GL.

The bending angles $\xi_r(p_r)$, $\xi_d(p_d)$ and main refractive part of the phase path excess $\kappa_r(p_r)$, $\kappa_d(p_d)$ are connected by the next relationships:

$$\xi_{d,r}(p_{d,r})=-p_{d,r}\left[\int_{r_0,\,a}^{R_G}\frac{dn}{ndr}\frac{dr}{\sqrt{n^2(r)r^2-p_{d,r}^2}}+\int_{r_0,\,a}^{R_L}\frac{dn}{ndr}\frac{dr}{\sqrt{n^2(r)r^2-p_{d,r}^2}}\right], \quad (6.1.8)$$

$$\kappa_{d,r}(p_{d,r})=-\left[\int_{r_0,\,a}^{R_G}\frac{dn}{ndr}\sqrt{n^2(r)r^2-p_{d,r}^2}\,dr+\int_{r_0,\,a}^{R_L}\frac{dn}{ndr}\sqrt{n^2(r)r^2-p_{d,r}^2}\,dr\right], \quad (6.1.9)$$

where r_0 is the minimal distance of the direct ray trajectory LG from the centre of spherical symmetry (point O). The lower limits r_0 and a in the integrals (6.1.8), (6.1.9) are corresponding to the direct and reflected signals, respectively. The next connection between the bending angle and the main refractive part of the phase path excess follows from Eqs. (6.1.8) and (6.1.9)

$$\xi_{r,d}(p_{r,d}) = -d\kappa_{r,d}(p_{r,d}) / dp_{r,d} \,. \tag{6.1.10}$$

Eqs. (6.1.5)–(6.1.10) describe the phase delays of the reflected and direct signals as functions of the impact parameters $p_{r,d}$ and the distances r_0 , a . The bending angles $\xi_r(p_r)$, $\xi_d(p_d)$ and the main refractive part of the phase path excess $\kappa_r(p_r)$, $\kappa_d(p_d)$ are depending also on the altitude profile $n(r)$. Parameters $n(R_L)$, $n(R_G)$ are supposed to be equal unity. From Eqs. (6.1.5)–(6.1.7) one may calculate by use of the relationship (3.4.10) the difference of the Doppler shifts F_r and F_d corresponding to the reflected and direct signals.

$$\frac{dS_d}{dt} = -\lambda F_d = (p_d - p_s)\{(1/d_{1s} + 1/d_{2s})dp_s/dt - (p_d^2 - p_s^{\,2})\{R_L dR_L/dt$$
$$[d_{1s}(d_{1s} + d_{1d})(p_d d_{1s} + p_s d_{1d})]^{-1} + R_G dR_G/dt[d_{2s}(d_{2s} + d_2)(pd_{2s} + p_s d_{2d})]^{-1}\}\}, \tag{6.1.11}$$

$$d_{1s} = (R_L^{\,2} - p_s^{\,2})^{1/2},\; d_{2s} = (R_G^{\,2} - p_s^{\,2})^{1/2},$$
$$d_{1d} = (R_L^{\,2} - p_d^{\,2})^{1/2},\; d_{2d} = (R_G^{\,2} - p_d^{\,2})^{1/2}, \tag{6.1.12}$$

$$\frac{dS_r}{dt} = -\lambda F_r = (p_r - p_s)\{(1/d_{1s} + 1/d_{2s})dp_s/dt - (p_r^2 - p_s^{\,2})\{R_L dR_L/dt$$
$$[d_{1s}(d_{1s} + d_{1r})(p_d d_{1s} + p_s d_{1r})]^{-1} + R_G dR_G/dt[d_{2s}(d_{2s} + d_{2r})(pd_{2s} + p_s d_{2r})]^{-1}\}\} - \tag{6.1.13}$$
$$-2\frac{d[n(a)a]}{n(a)adt}\sqrt{n^2(a)a^2 - p_r^2},$$

$$d_{1r} = (R_L^{\,2} - p_r^{\,2})^{1/2},\; d_{2d} = (R_G^{\,2} - p_r^{\,2})^{1/2}. \tag{6.1.14}$$

where λ is the wavelength relevant to the carrier frequency of the emitted signals. Eqs. (6.1.11) and (6.1.13) are valid for general case of the non-circular orbits of the GPS and LEO satellites. Eqs. (6.1.11) and (6.1.13) can be simplified under condition

$$\left| \begin{matrix} (p_{d,r}^2 - p_s^{\,2})\{R_L dR_L/dt[d_{1s}(d_{1s} + d_1)(p_{d,r}d_{1s} + p_s d_1)]^{-1} + \\ R_G dR_G/dt[d_{2s}(d_{2s} + d_2)(p_{d,r}d_{2s} + p_s d_2)]^{-1}\} \end{matrix} \right| << \tag{6.1.15}$$
$$<< \left|1/d_{1s} + 1/d_{2s})dp_s/dt_s\right| \;,$$

which is valid if $\left|(p_{d,r} - p_s)dR_{L,G}/dt\right| << p_s\left|dp_s/dt\right|$. This inequality is valid for all practical RO situations. Under indicated approximation one can obtain from Eqs. (6.1.11), (6.1.13)

$$\frac{dS_d}{dt} = -\lambda F_d = (p_d - p_s)(1/d_{1s} + 1/d_{2s})dp_s/dt \,, \tag{6.1.16}$$

$$\frac{dS_r}{dt} = -\lambda F_r = (p_r - p_s)(1/d_{1s} + 1/d_{2s})dp_s/dt - 2\frac{d[n(a)a]}{n(a)adt}\sqrt{n^2(a)a^2 - p_r^2} \,, \tag{6.1.17}$$

If the atmosphere is spherically symmetric and large-scale terrain features are absent, i.e. $a = \text{const}$ in the region of measurements, then expression (6.1.17) may be simplified

$$\frac{dS_r}{dt} = (p_r - p_s)(1/d_{1s} + 1/d_{2s})dp_s / dt . \qquad (6.1.18)$$

According to (6.1.16), (6.1.18) the difference of the Doppler shifts of the direct and reflected signals ΔF is equal to

$$\Delta F = F_d - F_r = -\lambda^{-1}(p_d - p_r)(1/d_{1s} + 1/d_{2s})dp_s / dt . \qquad (6.1.19)$$

Equation (6.1.19) may be transformed in the case of small elevation angle ψ to the form

$$p_d - n(a)a = h_d - [n(a) - n(h_d)]a \approx -\lambda \Delta F \frac{d_{1s} d_{2s}}{R_{GL}(dp_s / dt)} , \qquad (6.1.20)$$

where h_d is the height of the ray perigee relevant to the direct signal. The approximate Eq. (6.1.20) is fulfilled under condition $\psi^2 << 1$. In this case the minimal height h_d of direct ray may be estimated using measurements of the Doppler shifts corresponding to the reflected and direct signals.

Eqs. (6.1.5)–(6.1.9) and (6.1.19) may be used for evaluating the difference of the phase path delays and Doppler frequency of the direct and reflected signals. In addition to the phase delay and Doppler frequency the refractive attenuations $X(p_{d,r})$ of the direct and reflected signals may be evaluated [14,21,30,31,53]

$$X(p_{d,r}) = p_{d,r} R_{GL}^2 \left[\sqrt{R_L^2 - p_{d,r}^2} \sqrt{R_G^2 - p_{d,r}^2}) R_L R_G \right]^{-1} \left[\sin\theta \, \partial\theta / \partial p_{d,r} \right]^{-1}, \qquad (6.1.21)$$

where R_{GL} is the distance GL, $\partial\theta / \partial p_{d,r}$ are the partial derivative which are evaluated under condition of constancy of the distances R_L, R_G

$$\partial\theta / \partial p_d = d\xi_d(p_d) / dp_d - 1/\sqrt{R_L^2 - p_d^2} - 1/\sqrt{R_G^2 - p_d^2} , \qquad (6.1.22)$$

$$\partial\theta / \partial p_r = 2d\xi_r(p_r) / dp_r - 2d\psi(p_r) / dp_r - 1/\sqrt{R_L^2 - p_r^2} - 1/\sqrt{R_G^2 - p_r^2} . \qquad (6.1.23)$$

The refractive attenuation $X(p_r)$ in (6.1.21) corresponding to the reflected signal is equal to the product of the generalized coefficient of the spherical divergence D_E and the refractive attenuation in the atmosphere along the path 2 (Fig. 6.1.1). The spherical divergence D_E accounts for the size changes of ray tube owing to the specular reflection from the smooth spherical surface of the earth [30, 31]. Expressions (6.1.21)–(6.1.23) may be applied for calculating the refractive attenuation of the direct and reflected signals using the corresponding formulas (6.1.1), (6.1.2) which connect the angle θ with the impact parameters $p_{d,r}$. It is important to consider the partial case when the atmosphere is absent. For this case the formulas (6.1.1), (6.1.2), (6.1.4), (6.1.22), (6.1.23) may be transformed to relationships

$$\theta = \pi - \sin^{-1}[p_{d0} / R_L] - \sin^{-1}[p_{d0} / R_G], \tag{6.1.24}$$

$$\theta = \pi - 2\psi_0 - \sin^{-1}[p_{r0} / R_L] - \sin^{-1}[p_{r0} / R_G], \tag{6.1.25}$$

$$\cos\psi_0 = p_{r0} / a, \tag{6.1.26}$$

$$\partial\theta / \partial p_{d0} = -1 / \sqrt{R_L^2 - p_{d0}^2} - 1 / \sqrt{R_G^2 - p_{d0}^2}, \tag{6.1.27}$$

$$\partial\theta / \partial p_{r0} = -2 / \sqrt{a^2 - p_{r0}^2} - 1 / \sqrt{R_L^2 - p_{r0}^2} - 1 / \sqrt{R_G^2 - p_{r0}^2}, \tag{6.1.28}$$

$$S_{r0} = \sqrt{R_L^2 - p_{r0}^2} + \sqrt{R_G^2 - p_{r0}^2} - 2\sqrt{a^2 - p_{r0}^2}, \tag{6.1.29}$$

$$S_{d0} = \sqrt{R_L^2 - p_{d0}^2} + \sqrt{R_G^2 - p_{d0}^2}, \tag{6.1.30}$$

$$\Delta F_g = \lambda^{-1}(F_{d0} - F_{r0}) \approx -\lambda^{-1}(p_{d0} - p_{r0})(1/d_{1s} + 1/d_{2s}) dp_s / dt, \tag{6.1.31}$$

where ψ_0 is the grazing angle, $S_{r0,d0}$ are the phase path delays, ΔF_g is the difference of the Doppler shifts F_{d0}, F_{r0} relevant to the direct and reflected signals, p_{d0}, p_{r0} are the impact parameters corresponding to the trajectories of the reflected and direct signals in the case of free space.

Below we will consider an important relationship, which connects the derivative of the Doppler shifts with respect to time and the refractive attenuation of radio waves. If the atmosphere is spherically symmetric and and surface roughness is absent, under conditions

$$\left|p_{d,r} - p_s\right| << p_s \quad \left|(p_{d,r} - p_s) d(\partial\theta / \partial p_s) / dt\right| << \left|(dp_{d,r} / dt - dp_s / dt)\partial\theta / \partial p_s\right|, \tag{6.1.32}$$

one can obtain according to Eqs. (6.1.16) and (6.1.18) the next approximated relationships

$$\frac{dF_{d,r}}{dt} = -\left(\frac{dp_{d,r}}{dt} - \frac{dp_s}{dt}\right)\left(m\frac{dp_s}{dt}\right)^{-1}, \tag{6.1.33}$$

$$m = \frac{d_{1s} d_{2s}}{R_{GL}}\left(\frac{dp_s}{dt}\right)^{-2}, \tag{6.1.34}$$

$$\frac{d\Delta F}{dt} = -\left(m\lambda\frac{dp_s}{dt}\right)^{-1}\left(\frac{dp_d}{dt} - \frac{dp_r}{dt}\right), \tag{6.1.35}$$

where m is slowly changing function of time which may be determined from the satellites orbital data. By use of equation $dp/dt - dp_s/dt \approx [X(t) - 1] dp_s / dt$, [22], which is valid for direct and reflected signals, one can obtain from (6.1.33)–(6.1.35)

$$m\lambda \frac{dF_{d,r}}{dt} = \left(1 - X_{d,r}\right),\ X_{d,r} = 1 - m\lambda \frac{dF_{d,r}}{dt}, \tag{6.1.36}$$

$$\frac{d\Delta F}{dt} = -(m\lambda)^{-1}\left(X_d - X_r\right). \tag{6.1.37}$$

It follows from Eqs. (6.1.36) that the derivative of the Doppler frequencies $F_{d,r}$ with respect to time is proportional to the refractive attenuation variations of the direct and reflected signals, respectively. Usually value X_d is many times greater X_r. So in the free space when $X_d = 1$ the derivative of difference ΔF is nearly constant because m is a slowly changing parameter

$$\frac{d\Delta F}{dt} \approx -(m\lambda)^{-1}. \tag{6.1.38}$$

The second equation (6.1.36) allows computing the refractive attenuations $X_{d,r}$ directly from the phase and/or frequency data. This is useful for avoiding the refractive attenuation from the amplitude data and estimating the total absorption of radio waves at the paths 1 and 2 (Fig. 6.1.1). Value X_r accounts for the generalized spherical divergence factor and the refractive attenuation of the reflected waves along the path 2. This opens the way to measure the reflection coefficient of the earth's surface which includes the total absorption of radio wave along the ray path 2. According to theory of radio waves scattering [31, 168], the reflection coefficient of the Earth as a planet observed from outer space is related to parameter X_r as follows

$$\eta^2(\psi) = \frac{\chi_r X_r V^2(\psi)\Phi(\psi)}{X_d \chi_d}, \tag{6.1.39}$$

where χ_r, χ_d are the total absorption coefficients of radio waves on the paths 1 and 2; $\Phi(\psi)$ is the factor that accounts for influence of the large-scale terrain features, shadowing, and some other effects; and $V^2(\psi)$ is the reflection coefficient of a plane wave incident at a grazing angle ψ on an atmosphere-ground interface. The coefficient $V^2(\psi)$ depends on the permittivity ε of ground and polarization of radio waves. If emitted radio waves and receiving antenna are circularly polarized, the coefficient $V^2(\psi)$ is related to the grazing angle ψ and the permittivity ε as follows

$$\begin{gathered} V^2(\psi) = (s_g + s_v)^2/4,\ s_g = (A-1)/(A+1), \\ s_v = (\varepsilon A - 1)/(\varepsilon A + 1),\ A = \sin\psi/(\varepsilon - \cos^2\psi)^{1/2}. \end{gathered} \tag{6.1.40}$$

Values of the parameter ε relevant to the sea surface are known from experimental data [169]. The factor $\Phi(\psi)$ in the decimeter range is nearly equal to unity at grazing angles. Therefore equation (6.1.39) enables measurement of the total atmospheric absorption of radio waves from measurements of X_r and reflection coefficient η^2 from the phase and amplitude data, respectively. Below analysis of RO data obtained during MIR/GEO RO mission [30, 31, 170] will be provided.

6.1.3. Reflected signals in MIR/GEO RO measurements

Scheme of the MIR/GEO RO experiments [30, 31] is shown in Fig. 6.1.1 (left panel). Transmitter installed on the orbital station "MIR" (point L) emitted at wavelength $\lambda = 32$ and 2 cm monochromatic radio waves which have been registered by receivers installed on a geostationary satellite (point G) [29]. RO signals have been transmitted from geostationary satellites to a ground based station for analysis. Open-loop phase tracking method was used for removing Doppler effect due to orbital motion of the satellites and expected atmospheric influence. Then a spectral method was applied for revealing reflected signals in the RO data. The Doppler spectra of RO signal reflected from the surface of the Indian Ocean are shown in Fig. 6.1.1 (right panel, wavelength $\lambda = 32$ cm). The Doppler frequency (expressed in Hz) is shown at horizontal axis, the spectral intensity of reflected signal (normalized to the spectral intensity of direct signal) is given on vertical axis. The data are shown in succession with time interval between neighboring spectra equal to 1/3 s. As follows from Fig. 6.1.1 the distribution of spectral power density of reflected signal depends on time and may be connected with features in long-scale slope variations of the sea surface and physical properties of boundary layer of the atmosphere.
Reflected signal has been observed also at wavelength 2 cm during RO event 28 May, 1998. Time-Doppler frequency story of signal reflected from the Aral Sea is shown in Fig. 6.1.2. A weak reflected signal (sharp strip above, near frequency 20 KHz) is observed together with the more intense direct signal (broad bright strip below near frequency 25 KHz). Difference of Doppler frequencies of reflected and direct signal gradually diminishes as the time increases. Reflected signal disappeared after departure of a specular point from Aral Sea because roughness of the land surface.

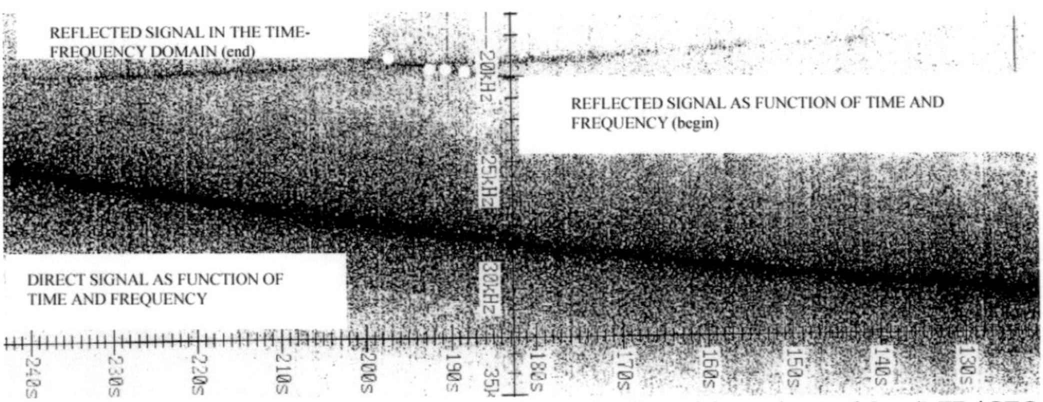

Fig. 6.1.2. The time-Doppler frequency story of signal reflected from the Aral Sea (MIR/GEO data, wavelength 2 cm).

6.1.4. Influence of boundary layer on parameters of reflected signal

In Fig. 6.1.3 (right panel) dependence of the Doppler frequencies difference $\Delta F(t)$ (Hz) of the direct and reflected signals is shown as a function of time t for RO event December 25, 1990.

The RO region located above the Indian Ocean at 31.2° S latitude and 68.2° E longitude. Curve 1 describes experimental data, curves 2, 3 correspond to two models of $\Delta F(t)$. The first model describes the function $\Delta F(t)$ for the case of free space propagation (curve 2). The second model shows influence of the real atmosphere with the near surface refractivity equal to N_0 = 350 N-units, and vertical gradient $dN(h)/dh$ = –36.8 (N-units/km) (curve 3). As follows from Fig. 6.1.3 (left panel) the atmospheric influence is essential near zone of radio shadow. Intensity of reflected signal has sharp changes near radio shadow region. Reflection coefficient varies from nominal level 0.02…0.04 up to 0.08…0.14 in the time interval 250…270 s and then indicates sharp changes from 0.03 up to 0.5. These variations reveal horizontal inhomogeneity of the atmosphere along the trajectory of motion of the specular reflection point. As follows from model [30,31] the reflection coefficient depends on the vertical gradient of refractivity in the boundary layer. Theoretical dependence of the refractive attenuation of reflected signal on the vertical gradient of refractivity is given in Fig. 6.1.4 (left panel). The central angle θ is shown at the horizontal axis. The refractive attenuation of the reflected signal is shown on the vertical axis in Fig. 6.1.4. Curve 1 corresponds to the case when the atmosphere is absent. Curve 2–5 are relevant to the vertical gradients equal to $dN(h)/dh =$ –35.4 N-units km^{-1} $dN_1(h)/dh = -43.0$; $dN_2(h)/dh = -50.0$ and $dN_3(h)/dh = -57.0$ N-units km^{-1}. The angular position of radio shadow is different for indicated values of vertical gradient $dN(h)/dh$ and changes from 102.3° (absence of the atmosphere, $dN(h)/dh = 0$) to 103.5° $(dN_1(h)/dh = -35.0)$, 103.75° $(dN_2(h)/dh = -43.0)$, 103.9° $(dN_3(h)/dh = -50.0)$, to 104.1° $((dN_4(h)/dh = -57.0)$.

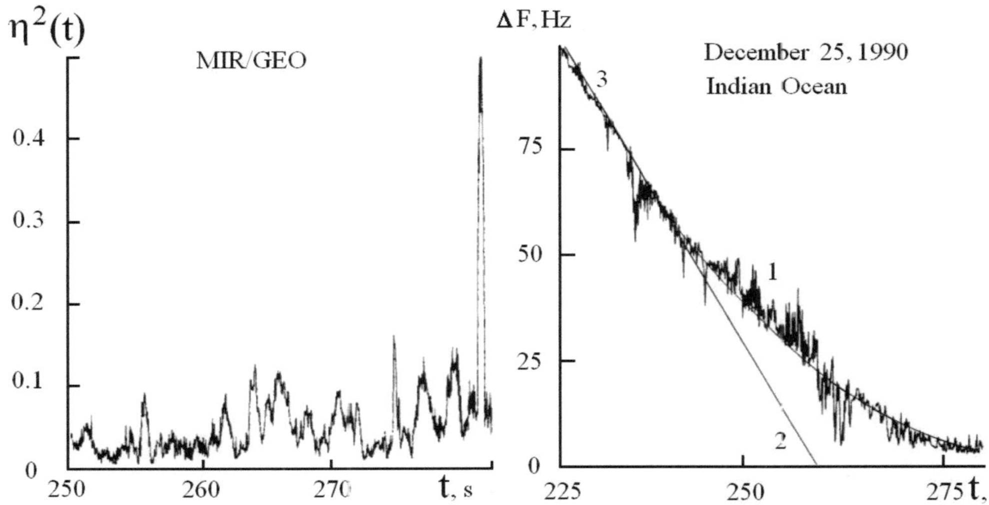

Fig. 6.1.3. Comparison of measured and theoretical values of the reflection coefficient $\eta^2(t)$ (left). Doppler frequencies difference $\Delta F(t)$ of the direct and reflected signals (right).

As follows from Fig. 6.1.4 the refractive attenuation of reflected signal near the boundary of radio shadow is a sharp function of the central angle θ and vertical gradient $dN(h)/dh$.

Thus the amplitude of reflected signal near the radio shadow is an indicator of vertical gradient of refractivity in the boundary layer of the atmosphere.

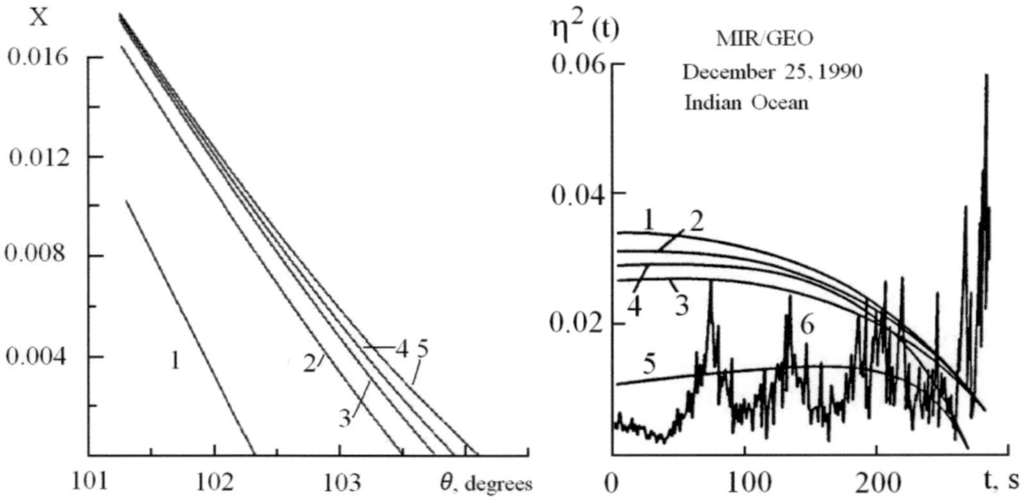

Fig. 6.1.4. Dependence of the refractive attenuation of reflected signal on the vertical gradient of refractivity (left). Comparison of the measured and theoretical values of the reflection coefficient $\eta^2(t)$ (right).

6.1.5. Estimation of the total absorption of radio wave from parameters of reflected signal

The following relationships for $\eta^2(\psi)$ are satisfied under condition $R_L \sin\psi << a$ [31]

$$\eta^2(\psi) = V^2(\psi)\,\chi_r\left[\partial\theta/\partial p_d\right]/\left[\chi_d\,\partial\theta/\partial p_r\right] \approx V^2(\psi)\,\chi_r X_d\,/(\chi_d X_r)\,. \qquad (6.1.41)$$

Values of the refractive attenuations X_d, X_r along the path 1 and 2 may be measured by use of the phase data as shown in section 6.1.2. Therefore measurements of the reflection coefficient $\eta^2(t)$ and knowledge of $V^2(\psi)$ enable estimation of the ratio of total absorptions χ_r / χ_d along the path 1 and 2 (Fig. 6.1.1). In the decimeter range the atmospheric absorption along the path 1 is small and the total absorption coefficient χ_d is near to unity. However, the radio waves propagates along the path 2 through all altitudes in the atmosphere at near grazing angles and are more sensitive to the absorption effect.

The RO measurement session was conducted at wavelength of $\lambda = 32$ cm during the MIR orbital station immersion into the Earth's radio shadow. The reflected wave was received at a geostationary satellite. The reflected and the direct signals were retransmitted from the geostationary satellite to an ground based receiving station in an analog format and have been recorded on a magnetic tape in an analog format. Then signals were converted to a digital format and processed by usage of trajectory data of the MIR orbital space station. Processing yielded the form and power of the Doppler spectrum of the reflected signal and its position relative to the direct signal as given by difference of the Doppler frequencies $\Delta F(t)$. The reflection coefficient $\eta^2(t)$ was computed from measured powers of the reflected and direct signals. The measured reflection coefficients $\eta^2(t)$ are shown and compared with their equivalents in Fig. 6.1.4 (right panel). Values of the experimental reflection coefficients are shown on vertical axis. The elapsed time from the session start is indicated on the horizontal axis in seconds. Curves 1-3 represent the numerically calculated reflection coefficient in presence of the atmosphere for N_0 = 320 N-units and $dN(h)/dh = -40.5\ N-\text{units km}^{-1}$. These magnitudes of N_0 and $dN(h)/dh$ agree with the standard refractive index profile in the atmosphere [61]. Curve 4 describes dependence $\eta^2(t)$ on time for the case when the atmosphere is absent. Curves 1, 2, 3, and 4 have been calculated for the sea-water conductivity of σ_W=2.7, 4.0; and 7.0 mhos/m, respectively. According to [169], the real part of the permittivity has been taken to be equal to 79. Curve 5 has been constructed for σ_W = 4 mhos/m, with accounting for the total absorption in atmospheric oxygen. Curve 6 is an experimental one. The time instant when the MIR orbital space station entered the geometrical radio shadow relative to the geostationary satellite corresponds to an intersection of curve 4 with the horizontal axis. The experimental reflection coefficient reached a maximum in the radio shadow zone and is subjected to significant variations. Outside the radio shadow zone, curve 5 satisfactorily fits in average the experimentally measured reflection coefficient $\eta^2(t)$. Comparison of curve 5 with curves 1-3 shows that the total absorption by the atmosphere over measurement site was equal to 4.8±1.2 dB. This corresponds to an average absorption coefficient of 0.0096±0.0024 dB/km for the path length of radio wave in the atmosphere equal to 500 km. This value closely agrees with the absorption in atmospheric oxygen as measured by the radio occultation technique in [29] and also with theoretical and experimental values reported in [128].

6.1.6. Radio image of boundary layer and reflective surface

For more careful analysis of connections of parameters of reflected signal with structure of the boundary layer it is useful to consider radio image of the troposphere and reflective surface. Radio image indicates multibeam propagation and resolves different rays propagating in the boundary layer. A radioholographic method has been applied to obtain radio image of the boundary layer and terrestrial surface [52, 73]. Examples of radio images of the Earth surface and boundary layer of the troposphere are shown in Fig. 6.1.5 and Fig. 6.1.6.

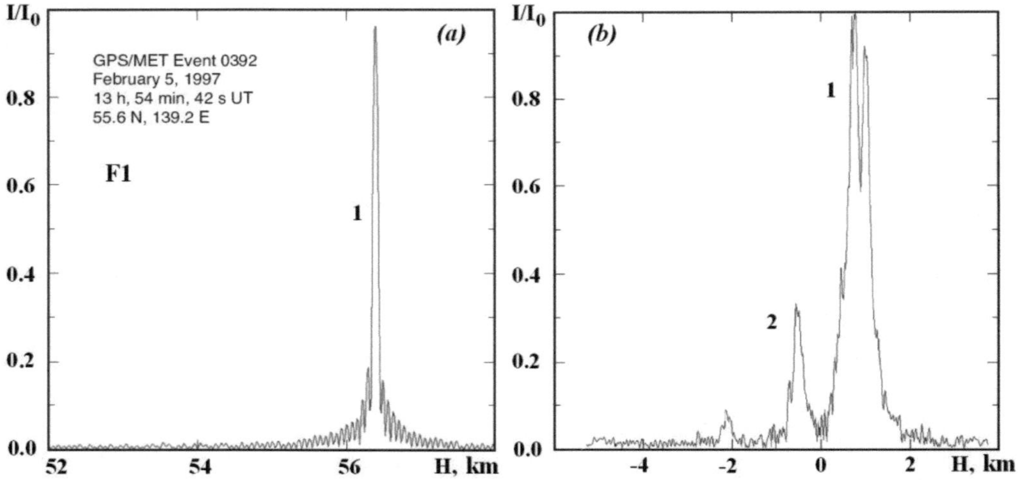

Fig. 6.1.5. Vertical distributions of radio brightness in the troposphere and signals reflected from the earth's surface at the first GPS frequency L1.

These radio images were obtained from GPS/MET RO data at two frequencies f_1 = 1575.42 MHz (Fig. 6.1.5) and f_2 = 1227.6 MHz (Fig. 6.1.6). The height along the visible Earth's limb as seen from the LEO satellite is shown along the horizontal axis. Negative values of height correspond to signals reflected from the Earth's surface. Fig. 6.1.5(a) shows a narrow (half-power width ~ 20 μ rad) single ray in the mesosphere at the 56 km altitude. This translates into the vertical resolution of about 60 m. Fig. 6.1.5(b) shows the complex rays structure about 1/2 minute later for the same occultation event. Here the perigee point of the direct ray lies deep in the lower troposphere between 1 and 2 km above sea level. This figure also includes a weaker signal corresponding to a near-specular reflection from the ocean. The similar examples of radio images of the Earth's surface and boundary layer of the troposphere at frequency f_2 are shown in Fig. 6.1.6. Fig. 6.1.5 and Fig. 6.1.6 allow following a time story of reflected signal. The reflected signal is separated from direct ray in the right panel of Fig. 6.1.5 and in left panel of Fig. 6.1.6. In right panel of Fig. 6.1.6 phenomena of conjunction of direct and reflected signal near a radio shadow zone is clearly seen. This phenomenon has been observed earlier in MIR/GEO data above the Indian Ocean [30]. Above the sea surface vertical gradient of the refractivity is diminishing because the humidity influence and therefore an effect of disruption of trajectory of reflected signal from the Earth's surface is arisen and one may observe its propagation in the boundary layer to direct ray. A theory describing conditions in the atmosphere for initialization of this phenomenon has been elaborated in [30, 31, 170]. It follows from this analysis, that the parameters of the reflected signal near radio shadow zone are the function of the meteorological parameters of the boundary layer.

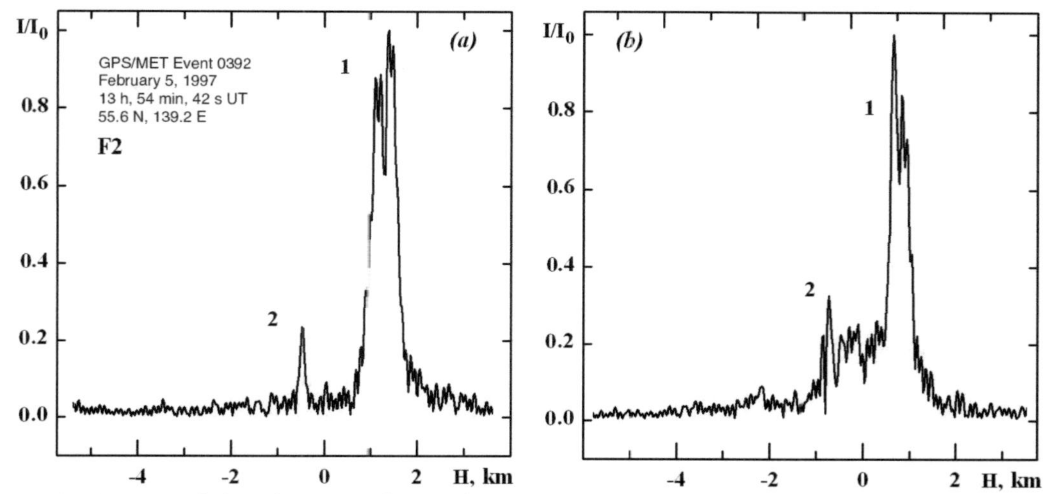

Fig. 6.1.6. Vertical distributions of radio brightness in the troposphere and signals reflected from the earth's surface at the second GPS frequency F2.

6.2. Absorption of radio waves and possibilities of monitoring of water vapor and clouds

Centimeter and millimeter radio waves experience attenuation in the troposphere, since the absorption lines of oxygen and water vapor are located in this range. For analysis of the radio wave attenuation the absorption coefficient γ was introduced, assuming, that decreasing of the power density by value dP is proportional to the elementary length of radio ray dx and the initial value of the power density of radio wave $dP = -\gamma P dx$. The total absorption, equal to the decrease of value P relative to the initial power P_0, is determined by the expression

$$Y = P/P_0 = \exp(-\int \gamma(h)\, dx). \qquad (6.2.1)$$

Integration is performed along the ray path in the interval of the atmospheric part of the communication link GL with accounting for dependence of the absorption coefficient γ on the height. Since troposphere contains two absorbing gases, the total absorption coefficient on an unit length is expressed in decibels, then $\gamma = \gamma_w + \gamma_0$, where γ_w and γ_0 are, respectively, the absorption coefficients of water vapor and oxygen.

The molecule of water vapor has a constant electrical torque which interacts with electromagnetic field and causes the absorption of radio waves. Three lines of the resonance absorption are located at frequencies 22.3 GHz, 183.4 GHz and 323.8 GHz. The theory of the radio wave absorption by water vapor was described in [171, 172], comparison of the theory with experiments revealed a distinction and this required specification of the parameters of absorption lines. As it turned out it is necessary to account for the contributions of absorption, caused by the wings of the remote spectral lines [173, 174] . Multiple measurements of the absorption of millimeter radio waves by water vapor were accomplished. These experiments showed that, in accordance with the theory, is proportional to the absolute humidity and the square of pressure. The temperature dependence is more complex, it can be approximately proportional to function T^{-1}. Therefore the absorption coefficient by water vapor can be given by the approximate formula

$$\gamma_{\mathrm{w}} = \mathrm{w}_{\mathrm{a}} \, P_a^2 \, T^{-1} \, F_1(\lambda) \, . \qquad (6.2.2)$$

This presentation is convenient, because the meteorological parameters: the humidity w_a, pressure P_a and temperature T, as functions, depending on the height, are marked out as separate factors. The effect caused by the form of the absorption line required the quantum mechanical analysis is described by the function $F_1(\lambda)$. It is necessary to note that this factor selection is approximated, since the form of the spectral line also depends on the pressure and temperature. For the near surface's conditions $P_a = 1$ atm, T = 15°-20°C dependence of the absorption coefficient of the water vapor on frequency for $f < 200$ GHz is determined by the approximated formula

$$\gamma_{\mathrm{w}} = \left[0.067 + \frac{\alpha_1}{(f-22.3)^2 + \beta_1} + \frac{\alpha_2}{(f-183.4)^2 + \beta_2} \right] f^2 \, \mathrm{w}_a \cdot 10^{-4} \, . \qquad (6.2.3)$$

The physical values in (6.2.3) have the next dimensions: γ – dB·km^{-1}, frequency f – GHz, the humidity w_a – g·m^{-3}. The parameters in this formula for temperature equal to 15° and 20°C have, respectively, the next values: $\alpha_1 = 3$ and 2.4; $\alpha_2 = 10$ and 7.3; $\beta_1 = 7.3$ and 6.6; $\beta_2 = 9$ and 5. The results of calculations (solid curve) and measurements (data points) of the absortivity coefficient of the radio wave by water vapor near the earth's surface when $h = 0$, $P_a = 1$ atm, $T = 293°$ K, $\mathrm{w}_a = 7.5$ g·m^{-3} are shown in Fig. 6.2.1. In the papers [175-178] the results of evaluation of the millimeter radio wave's absorption by the water vapor are shown for different values of pressure and temperature.

Molecule of oxygen has the magnetic dipole momentum that causes appearance of the single absorption line at the frequency 119 GHz and a group lines at frequency 60 GHz. When the pressure is near one atmosphere the lines of this group resolved and looks like one broad line; the resolution of these line may be achieved at the altitudes greater 15 km. The theory of radio wave's absorption by oxygen is described in [179, 180]. With aim to determine more accurately the absorption coefficient by the oxygen many efforts were applied for evaluation of the spectral width of these lines, their relative intensity, their dependences on the pressure and height [181-186]. The absorption coefficient of oxygen depends on the pressure and temperature; this dependence can be approximated by formula

$$\gamma_{\mathrm{o}} = P_a^2 \, T^{-5/2} \, F_2(\lambda) \, . \qquad (6.2.4)$$

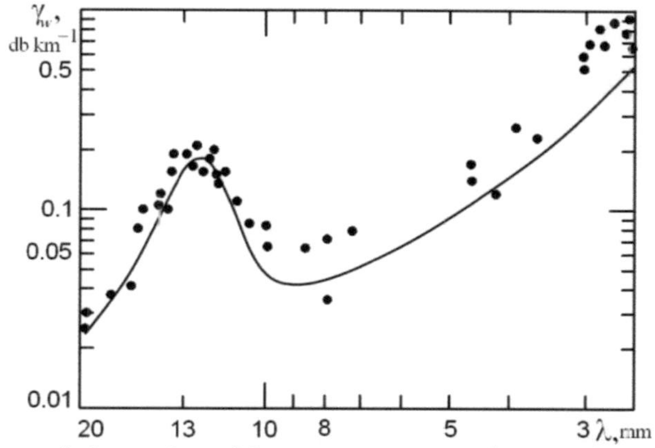

Fig. 6.2.1. Dependence of absorption of the water vapor on the wavelength.

Function $F_2(\lambda)$ near the resonance line of absorption is also dependent on the pressure and temperature. At frequencies $f < 56$ GHz the absorption coefficient for oxygen is given by the formula

$$\gamma_o = \left[\frac{6,6}{f^2 + 0.33} + \frac{9}{(f-57)^2 + 1.96}\right] f^2 \cdot 10^{-3} . \tag{6.2.5}$$

This formula corresponds to $P_a = 1$ atm, $T = 20°C$ and includes frequency expressed in GHz, and γ_0 given in dB·km^{-1}. dependence of the absorption coefficient of oxygen on upon wavelenght λ is shown in Fig. 6.2.2 for indicated values of P and T. The line in this Figure correspond to the theoretical, and the points give the experimental values of the absorption coefficient. Absorption of centimeter and decimeter waves by oxygen is carefully analyzed in paper [128]. From data of this paper the absorption coefficient γ_0 has next values when $h = 0$: for $\lambda = 21$-26 cm $\gamma_0 = -$ 0 016 dB·km^{-1}, when $\lambda = 16$-19 cm – 0.013. In the range $\lambda = 3$-15 cm – 0.012, and for $\lambda = 1.8$-2.5 cm – 0.015 dB·km^{-1}. The error in the experimental determination of γ_0 is about ± 0.003 dB·km^{-1}.

The short description of the absorption coefficients as function of the altitude is valid under the standard conditions, corresponding to the background values of the pressure, temperature and humidity near the earth' surface. In the problem of RO sounding it is necessary to determine the attenuation, corresponding to the radio ray LG, passing through the troposphere. This requires accounting for the changes in pressure, temperature and humidity with height. If one excludes from the consideration the centers of absorption lines having the large attenuation, then it is possible to provide the simple analysis of the problem. In [187, 188], it was indicated, that, if one accounts for the altitude dependences of humidity and temperature, then the altitude profiles of the oxygen and water vapor absorption coefficients are described by nearly exponential law

$$\gamma_w = \gamma_w(0) \exp\left(-hH_w^{-1}\right),$$
$$\gamma_0 = \gamma_0(0) \exp\left(-hH_0^{-1}\right). \tag{6.2.6}$$

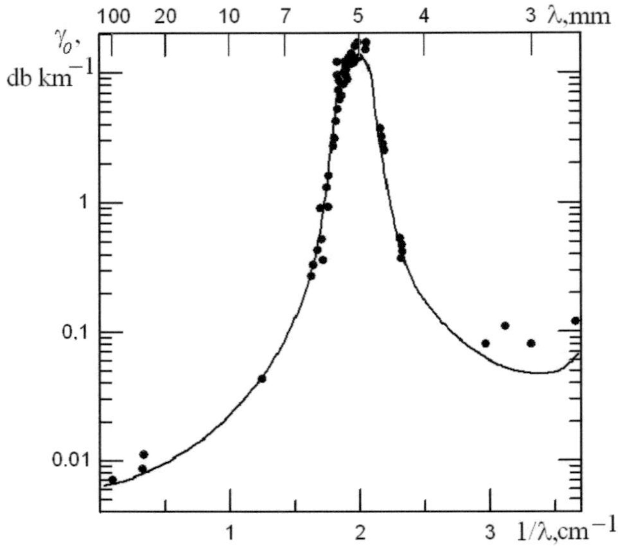

Fig. 6.2.2. Absorption coefficient of atmospheric oxygen.

Here H_w and H_0 are parameters, characterizing the diminishing of the absorption when the height increases, and $\gamma_w(0)$ and $\gamma_0(0)$ are correspond to $h = 0$, and can be found from formulas (6.2.3) and (6.2.5). dependence $\gamma_w(0)$ and $\gamma_0(0)$ on the wavelength is shown in Figs. 6.2.1 and 6.2.2. The total absorption of the radio wave along the radio ray LG is determined by the sum $Y = Y_w + Y_0$, where

$$Y_w = \exp\left[-\gamma_w(0) \int \exp\left(-hH_w^{-1}\right) dx\right],$$
$$Y_0 = \exp\left[-\gamma_0(0) \int \exp\left(-hH_0^{-1}\right) dx\right]. \quad (6.2.7)$$

During evaluation Y one may fulfilled integration along the straight line $T\mathrm{x}$, and the influence of refraction may be accounted for only for determination of the height of the radio ray perigee T-H (Fig. 4.3.1). In Fig. 4.3.1 the arbitrary point on the radio ray is located at the altitude h. Since the altitudes in the atmosphere are many times smaller the earth' radius $h << a$, then from Eq. (6.2.7) it follows

$$Y_w = \gamma_w(0)\left(2\pi\, a\, H_w\right)^{1/2} \exp\left(-H\, H_w^{-1}\right), \quad (6.2.8)$$

$$Y_o = \gamma_o(0)\left(2\pi\, a\, H_o\right)^{1/2} \exp\left(-H\, H_o^{-1}\right). \quad (6.2.9)$$

Eqs. (6.2.8) and (6.2.9) give dependences of the total absorption, caused by influence of oxygen and water vapor on the minimal height of radio ray H. It follows from (6.2.8) and (6.2.9), that the altitude dependence $Y(H)$ is determined by the parameters H_w and H_0 [187-191].

Absorption by oxygen is stable, it experiences insignificant annual changes and parameter H_0 is concentrated in the interval 4.3-5.1 km. In the broad range of decimeter and centimeter waves from $\lambda = 19$ cm up to $\lambda = 2$ cm one may consider γ_0 as a constant parameter which is equal to 0.014 dB·km^{-1}, therefore the total absorption by oxygen is expressed by a simple formula

$$Y_0(H) = 6.2 \exp(-0.2\, H), \tag{6.2.9A}$$

where Y_0 and H have dimensions dB and km, respectively. It follows from formula 6.2.9A, that for the centimeter waves near the earth's surface $Y_0 = 6.2$ dB, for $H = 5$ km one has $Y_0 = 2.3$ dB, and when $H >$ km the absorption due to oxygen is negligibly small.

It is necessary to account for the altitude dependence of the refractive attenuation $X(H)$ during determination from the RO data the absorption of decimeter waves caused by influence of oxygen. The results of determination of the oxygen absorption Y_0 at $\lambda = 32$ cm are published in [192]. The authors analyzed the data of five RO sessions conducted above the Indian Ocean, accounted for the theoretical dependence $X(H)$ and found the absorption $Y_0 = 5.5 \pm 1$ dB when $h = 0$, and $Y_0 \approx 2$ dB at the altitude $H = 4\text{-}5$ km. These data are well corresponding to the magnitudes calculated from (6.2.9A).

The usage of the absorption effect of centimeter and millimeter radio waves is necessary to control the distributions of water vapor and water content in the clouds. A preliminary analysis of the radio wave's absorption was conducted in [21, 193]. According to this analysis the large absorption of millimeter radio waves may make difficult the realizations of radio occultation experiments in the troposphere. Analysis, conducted in [21, 193], indicated, that the centimeter radio waves experience less absorption, this makes perspective their usage for monitoring of the distribution of water vapor in the clouds [194]. The water vapor absorptiones determined the water content; therefore it depends on seasonal and local meteorological conditions. A careful analysis of the radio waves absorption by water vapor is conducted in [177, 190], where values of the parameters H_w and $\gamma_w(0)$ have been estimated. From data of these papers for the moderate latitudes, when the wavelength is changing from 3 cm up to 0.8 cm in winter $H_w = 2.5$ km, and in summer $H_w = 1.8$ km. This parameter is proportional to the integral water content of the atmosphere; it may change from 0.8 km up to 2.9 km. At wavelength $\lambda = 1.35$ cm corresponding to the resonance absorption of water vapor the attenuation of radio waves is especially large. The standard profile of meteorological parameters was used in [21, 193] for evaluation of the altitude distribution of the absorption coefficient $\gamma_w(h)$, calculation was fulfilled for the average climatic conditions, typical for 60° Northern latitude in July (temperature near the earth' surface T=287 K, the pressure P=760 mm, the humidity w=11.1 g·m^{-3}) and in January (T=257 K, P=760 mm, w=2.1 g·m^{-3}). The altitude distribution of water vapor corresponds to a model of «humid» stratosphere: from surface up to tropopause – monotonic diminishing, above the tropopause up to 30 km– slow growing; the temperature is also decreases with height up to the tropopause, and then increases. The absorption coefficient of the molecular oxygen was evaluated from a quantum mechanical formula given in [183], its dependence on the temperature and pressure has been accounted for. The absorption coefficient of the water vapor has been evaluated from an engineer method, described in [176]. The results obtained by use of this method coincide with the experimental data for the standard conditions in the atmosphere. The results of evaluation of the total absorption Y for $\lambda = 1.35$ cm are presented in Fig. 6.2.3; curves 1 and 2 correspond to the absorption coefficient relevant to January and July. The relative contributions of the humidity and oxygen in the absorption coefficient depend on time and on the wavelength. At $\lambda = 1.35$ cm in summer the absorption in oxygen may be neglected, however in the winter time, when the humidity is diminishing, the contribution of oxygen into the absorption may be compared with the contribution of water vapor. The absorption is essentially greater in summer, than in winter, this is connected with the humidity change and indicates the high sensitivity of dependence $Y(H)$ to the distribution of water vapor in the atmosphere. For evaluation of the absorption $Y(H)$ the model of «humid» stratosphere has been used. This model accounts for an increase of the water vapor content above the tropopause; this explains no monotonous character of curves 1 and 2 with presence of sagging.

It follows from Fig. 6.2.3 that the RO technique is sensitive to vertical distribution of humidity: if suppose that the error in the amplitude measurements does not be greater than 1 dB, then the water content may determined at wave length $\lambda = 1.35$ cm in the range 20 to 30 km altitude. For $\lambda >> 1.35$ cm the absorption is diminishing, it is possible to conduct the experimental determination of dependence $Y(H)$ at lower altitudes for different meteorological conditions.

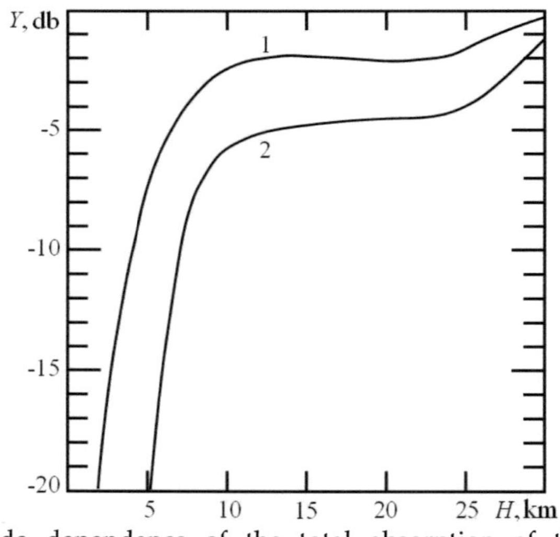

Fig. 6.2.3. The altitude dependence of the total absorption of the water vapor and atmospheric oxygen for $\lambda = 1.35$ cm.

The experimental investigations of absorption were accomplished in the communication link MIR -geostationary satellites [194]. For determination of absorption it is necessary to obtain the experimental dependence of the field strength of radio wave on the height of radio ray perigee $E(H)$, and then exclude the refractive attenuation. For excluding of the refractive attenuation the results of simultaneous measurements of the field strength E_1 of decimeter $\lambda_1 = 32$ cm and E_2 of centimeter $\lambda_2 = 2$ cm radio waves were used. Since the refractive attenuation is independent on the wavelength, the ratio E_2 / E_1 determined the additional total absorption at the centimeter waves caused by the water vapor under assumption, that the absorption of decimeter and centimeter waves by the oxygen is the same. For obtaining of the reliable altitude dependence of the total absorption it is necessary to use the results of several measurements of the altitude profile $Y(H)$ above different regions, the results of the independent measurements have been averaged in the interval $\Delta H \approx 0.2$ km. In Fig. 6.2.4A the points are indicating values of absorption of centimeter radio waves Y_w, obtained during RO measurements above the Pacific Ocean near the Aleutian Islands conducted on December 22 and 28, 1991. dependence of the total absorption of centimeter radio waves on the minimal height $Y_w(H)$, obtained from measurements above the Kazakhstan conducted on June 28 and July 02, 1991, are shown in Fig. 6.2.4B. Despite different regions, the difference in climatic conditions and seasons, the average values of the total absorption at the same altitudes are distinguishing insignificantly: at $H = 0$ in the region of the Aleutian Islands $Y_w = 14\pm3$ dB, and above the Kazakhstan $Y_w = 12\pm3$ dB. The consistence of the boundary layer of the atmosphere located near the earth's surface exerts significant influence on the spread of values Y_w at $H = 0$ km. More accurate values of absorption of centimeter radio waves were determined at the altitude $H = 0.4$ km; at this altitude in the region of the Aleutian Islands $Y_w = 11\pm2$ dB, and above Kazakhstan $Y_w = 9\pm2$ dB. Analysis of the results of eight RO easurements sessions above the sea and continental regions indicated, that in the altitude interval from 4 up to 5 km the absorption Y_w in average is equal to 3.8±1 dB.

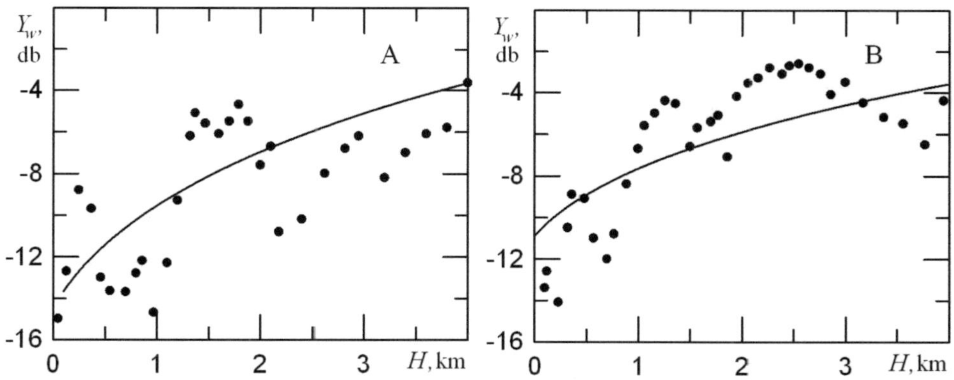

Fig.6.2.4. Total absorption of radio waves ($\lambda = 2$ cm) in different regions.

Analysis of two RO experiments, accomplished on December 25 and 26, 1991, above the Indian Ocean indicated that at $H = 0$ the absorption $Y_w = 11.5\pm2$ dB, and at $H = 4$ km it is equal to 3±1 dB. In the height dependence $Y_w(H)$ in the altitude interval from 1.5 up to 2.5

km a feature - a significant rise of the attenuation, caused by the absorption of radio waves in the clouds - has been observed. The authors of publication [194] used the found dependences $Y_w(H)$ and formula (6.2.8) for comparison of the theory and experimental data and obtained, that at $H_w = 1.5$ km and $\gamma_w(0) = (1.6\pm0.3)\cdot10^{-2}$ dB·km^{-1} the experimental data are corresponding to results of evaluation from (4.6.8). The curves in Fig. 6.2.4 found from the formula (6.2.8) under the indicated values of $\gamma_w(0)$ and H_w, showed the satisfactory correspondence with the experimental and theoretically evaluated dependences of $Y_w(H)$. The exponential altitude dependence of the total absorption of radio wave by water vapor is based on the assumption about of the exponential diminishing of the water vapor concentration with increasing of the altitude in the troposphere. This assumption in average corresponds to the practice; however the altitude humidity distribution for a local region depends on the season and meteorological conditions. The $\gamma_w(0)$ values obtained in the winter above the sea and in the summer period above the continental regions, are distinguishing by two times; in winter above the sea the absorption is greater, than in summer above the land. With accounting for the meteorological data about of the specific humidity in measurement regions the normalized value of the absorption coefficient of water vapor was determined: $\gamma_w / w_a = (2.2\pm0.4)\cdot10^{-3}$ dB·km^{-1}m^3·g^{-1}. This value corresponds to the results of the radiometric determination of absorption at wavelength $\lambda = 2$ cm published in [128]. By use of this value γ_w / w_a one can determine from measurement of Y_w the value of the absolute humidity w_a; from the data of RO measurements conducted near the Aleutian Islands $w_a = 14$ g·m^{-3}.

A theoretical analysis of the possible monitoring of the water vapor vertical distribution with usage of the dual frequencie RO measurements is given in [195]. The authors showed that the usage ranges $\lambda_1 \approx 1.75$ cm and $\lambda_2 \approx 1.48$ cm allows measuring vertical dependence of the integral water vapor absorption $Y_w(H)$ and finding from these data the altitude distribution of the atmospheric water content $w_a(h)$. The usage of the ratio Y_2 / Y_1 relevant to the ranges $\lambda_{1,2}$ excludes the interfering influence of the refractive attenuation, radio wave fluctuations and absorption by the oxygen.

Analysis of the radio waves absorption by the clouds is more difficult, since their types, water content, altitude and horizontal distribution distinguish very significantly . Meteorologists distinguish ten basic forms of the clouds. These forms are separated in many different kinds of the clouds structures. The most often encountered forms– the stratus (layered) clouds. These clouds are relatively homogeneous, have large horizontal size and a clearly determined vertical stratification . Meteorologists distinguish the stratus clouds of the lower level (0.5-1.8 km), moderate level (2-3.5 km height interval) and the upper level (with the height more than 4 km). For analysis of the absorption accounting for the contribution from the clouds of the lower and moderate level is important, where the stratus clouds have a liquid-drop structure. The upper level of the stratus clouds is located in region with negative temperature, where the water exists in the ice form; this part of the stratus clouds does not give any notable contribution in the radio waves absorption. The attenuation of the centimeter radio waves may be significant during radio occultation investigation of the extended in the horizontal direction stratus clouds of the lower level, and in the equatorial regions essential contribution will be obtained from the clouds of moderate level.

Let us consider two models of vertical distribution of the absorption coefficient $\gamma_c(h)$ of the layered clouds. The first model contemplates, that above and below a layered cloud the absorption is absent, and inside the cloud between the altitude h_1 (the upper boundary) and h_2 (the lower boundary) the absorption coefficient has the constant value γ_1. The second model described the absorption coefficient, $\gamma_c(h)$ by use of a relationship

$$\gamma_{\tilde{n}}(h) = \gamma_1 \exp\left[-\alpha_3^2 \left(h - h_3\right)^2\right], \tag{6.2.10}$$

where γ_1 is the absorption coefficient in the maximum of water content in a layered cloud, h_3 is the altitude of this maximum, and parameter α_3 determines the altitude velocity diminution of the water content. The absorption in the stratus clouds Y_c is determined by integration of the absorption coefficient $\gamma_c(h)$ along the axis x (Fig. 4.7.1), where $x = 0$ locates in the point T

$$Y_c(H) = 2\int_0^{x_m} \gamma_c(h)\, dx\,, \tag{6.2.11}$$

where

$$x_m = \left[\, 2a\,\left(h_m - H\right)\right]^{1/2}, \tag{6.2.12}$$

$$h = H + x^2/(2a)\,. \tag{6.2.13}$$

Here a is the Earth radius. It is accounted for in the Eqs. (6.2.11-6.2.13) that $a >> h_m$, where $h_m \approx 10$ km is the altitude in the troposphere, above which the absorption in the cloud may be neglected. For the first model of a homogeneous cloud with the sharp upper and lower boundaries it follows from Eq. (6.2.11)

$$Y_c\left(H\right) = 0 \qquad \text{when } H > h_1\,,$$

$$Y_c\left(H\right) = 2\gamma_1\left[\,2a\,\left(h_1 - H\right)\right]^{1/2} \qquad \text{when } h_2<H<h_1\,, \tag{6.2.14}$$

$$Y_c\left(H\right) = 2\gamma_1\left\{\left[\,2a\,\left(h_1 - H\right)\right]^{1/2} - \left[\,2a\,\left(h_2 - H\right)\right]^{1/2}\right\} \quad \text{when } H<h_2.$$

For the second model from Eqs. (6.2.10) and (6.2.11), with accounting for Eq. (6.2.13), one can obtain

$$Y_c\left(H\right) = 2\gamma_1\int_0^{y_m} \exp\left[-\alpha_3^2\left(H - h_3 + \frac{y^2}{2a}\right)^2\right] \mathrm{dy}\,. \tag{6.2.15}$$

The calculated altitude dependence of absorption $Y_c(H)$ is presented in Fig. 6.2.5 for two models of the clouds: curves 1 and 2 correspond to the first and second model, respectively, at wavelength $\lambda = 2$ cm. For evaluation of $Y_c(H)$ the next values of parameters were chosen: $h_1 = 2.4$ km, $h_2 = 1.9$ km, for the first model and $h_3 = 2.15$ km for the second model. Parameter γ_1 for both models was equal to 0.1 dB·km$^{-1,}$ and value $\alpha_3 = 3.54$ km^{-1}.
It follows from Fig. 6.2.5 that both modes give nearly the same dependences $Y_c(H)$: the maximal absorption is about of 13-16 dB at $H \approx 1.9$ km, and at the altitudes $H = 0.5$ km absorption is near 4 dB. It follows from Fig. 6.2.5 that in the communication link satellite-to-satellite the significant absorption of the centimeter waves should be observed with a typical sharp increasing of Y_c at the altitude H near the upper boundary of the cloud and decreasing in absorption at the altitudes H smaller than the height of lower boundary of a stratus cloud. In practice the layered clouds have a complex vertical profile, often with two and rare with three maxima in water content, therefore dependence $Y_c(H)$ may have also several maximum of absorption at different altitudes.

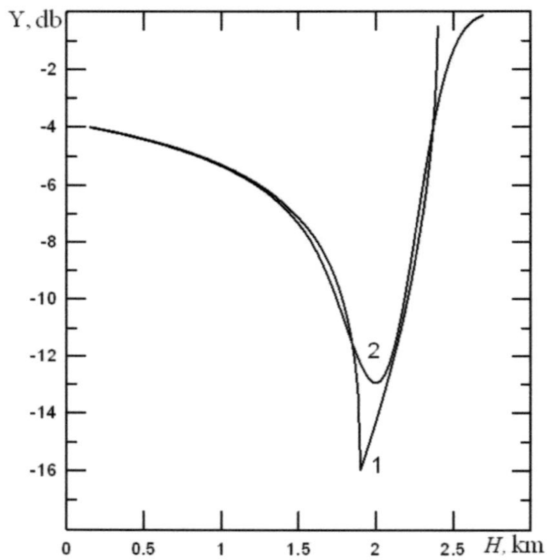

Fig.6.2.5. Altitude dependence of absorption of radio waves for two models of the stratus clouds; $\lambda = 2$ cm.

Let us compare the results of the absorption measurement during radio occultation of the clouds along the radio ray satellite to satellite, Y_c and during propagation of radio wave in vertical direction Y_v. In the case of measurements in vertical direction, for example, by an ground based radiometric device oriented to zenith, the absorption relevant to the first model the clouds is given by equation

$$Y_c = \gamma_1 \left(h_1 - h_2\right). \tag{6.2.16}$$

The same value for the second model will be

$$Y_c = \gamma_1 \pi^{1/2} \alpha_3^{-1}. \tag{6.2.17}$$

It follows from Eqs. (6.2.16) and (6.2.17), that for the indicated values of the absorption parameters Y_c relevant to vertical propagation of radio wave are very weak and are equal to 0.05 dB for both model. However in this case in the communication link satellite-to-satellite the large values of absorption will be observed. The combined contribution of the water vapor and clouds absorption at $\lambda = 2$ cm is presented in Fig. 6.2.6. As follows from this Figure at the altitudes $H = 4$ km, in the absence of the cloud's influence a weak absorption by water vapor should be observed equal to 2 dB. In the maximum of absorption at the altitude $H =$ 2 km the attenuation will achieve 17 dB, and before immersion of a satellite into the earth's limb, at the altitude $H = 0.5$ km, the attenuation will be equal to $Y = 11$ dB.

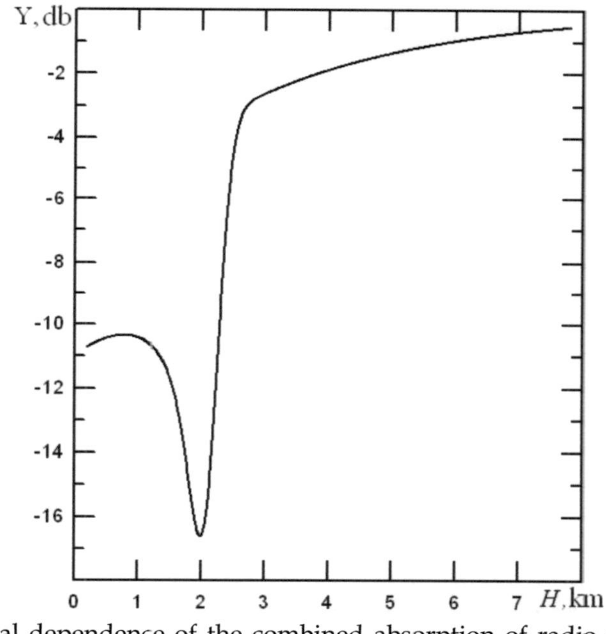

Fig.6.2.6. Theoretical dependence of the combined absorption of radio wave by the water vapor and clouds.

For determination of the altitude dependences of the radio wave attenuation presented in Fig. 6.2.5 and 6.2.6 values of the cloud absorption parameters, obtained in papers [128, 177] from data of the ground based measurements, were used. The corresponding information for the centimeter radio waves has insufficient accuracy; therefore the Fig. 6.2.5 and 6.2.6 give only approximated dependences $Y(H)$. During determination of the experimental dependences $Y(H)$ from data of RO measurements in the communication link station MIR-geostationary satellite it was necessary to diminish the interfering influence of the amplitude fluctuations and exclude the contributions of the refractive attenuation in the atmosphere and absorption in the atmospheric oxygen. To diminish the influence of the quick amplitude fluctuations the averaging of the experimental data for the time interval $\Delta t = 0.5$ c was fulfilled. Values of the refractive attenuation $X(H)$ and absorption in the oxygen $Y_0(H)$ were excluded after averaging from the experimental data. Therefore dependences $Y(H)$ for 64 of RO measurements in different regions were obtained. These dependences were separated in two groups: in the first group the observations with a monotonic growing of attenuation with corresponding decreasing of the altitude H were included, and into the second group vertical profiles with a quick increase of the attenuation in a sharp interval of the altitude H and the next decreasing of the attenuation below the height H. The typical examples of dependences $Y(H)$ from the first group are shown in Fig. 6.2.7. Curve 1 was obtained on 29 October 1997 for the region with latitudes 50° S, longitude 59° E; the curve 2 was obtained 09 July 1998 in the region with coordinate 43° S, 62° E. All these data do not indicate a cloud's influence; they may be considered as the indicators of absorption caused, mainly, by water vapor. From data in Fig. 6.2.7 at $H = 0.5$ km Y is equal to 8±1 dB, and at $H = 3$ km absorption is equal to 5±1 dB.

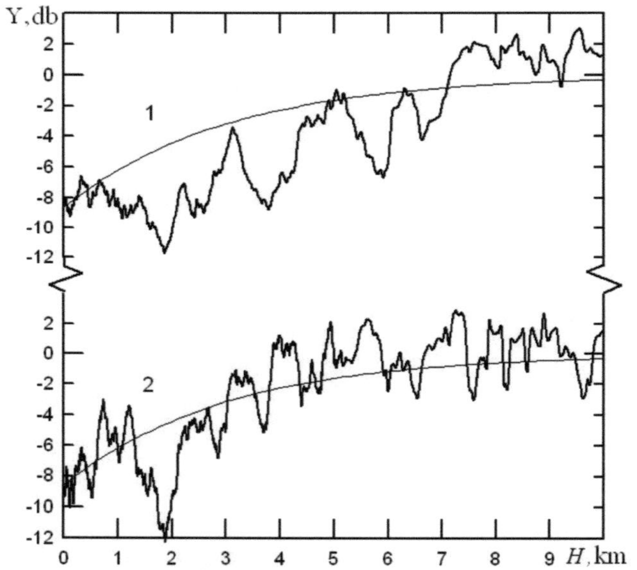

Fig.6.2.7. Experimental dependences of the radio wave absorption ($\lambda = 2$ cm) relevant to the first group of data.

From comparison of these data with the results of calculation (smooth curves) it is seen, that the registered attenuation is by 1.5 dB greater, than the expected theoretical values, if one suppose, that absorption is caused only by water vapor. This distinction of theory and experiment is located in the interval of measurement errors. In the first group data there are dependences $Y(H)$ having a regular increase in attenuation when the altitude H is decreasing, however it is difficult to explain this dependence only by influence of water vapor. The examples of these experimental dependences $Y(H)$ are shown in Fig. 6.2.8; the smooth curves indicate the theoretical dependence of the radio wave attenuation owing to the water vapor $Y_w(H)$. Curve 1 in this Figure corresponds to measurements fulfilled on 30 October 96 in the region with coordinates 19° N, 175° E, dependence 2 was obtained on 05 October 1996 in the region with coordinates 50° S, 56° E. It follows from comparison of the experimental dependences $Y(H)$ and theoretical $Y_w(H)$ that the additional attenuation is equal to 7±1 dB at the altitude $H = 0.5$ km. Effective length of a communication link satellite to satellite in the troposphere is near 200 km, therefore a possibility of realization of the scheme with the cloudless atmosphere is small. It seems that there are the contribution from the clouds in the data shown in Fig. 6.2.8. Exceeding of the measured values Y above the calculated values Y_w is observed for the most data in the first group.

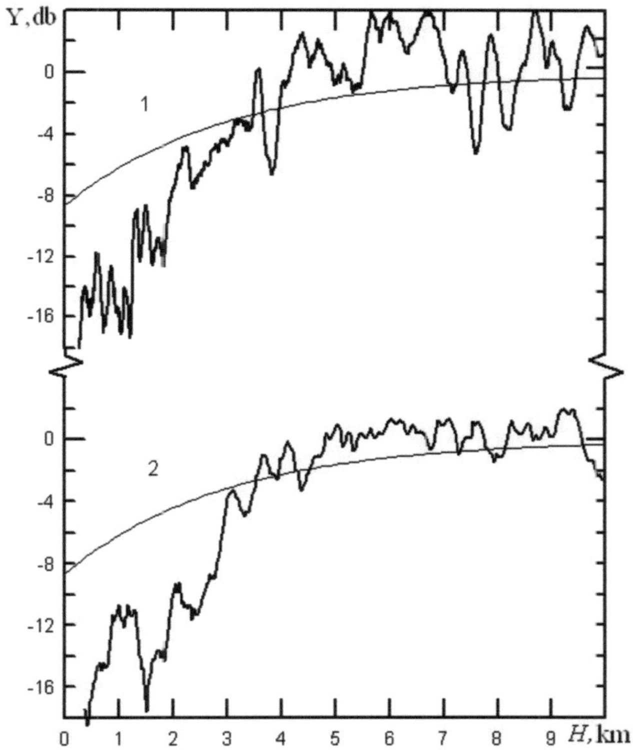

Fig.6.2.8. The typical altitude dependences of absorption $Y(H)$.

Examples of dependences $Y(H)$ belonging to the data of the second group are shown in Fig. 6.2.9, where once again become apparent the cloud's influence; curve 1 was obtained on 31 October 1997 for region with latitudes 51° S, longitude 59° E; dependence 2 was obtained on 05.06.1996, in the regions with coordinates 33° S, 173° E. The upper boundary of the clouds at the altitudes 2.5 and 3 km, the large absorption in the altitude interval $\Delta H \approx$ 1-2 km, contained in the interval 18-22 dB, and absorption's decreasing at the altitudes H=0.5 km up to 12±2 dB are seen in Fig. 6.2.9. The correspondence of the experimental data in Fig. 6.2.9 with the theoretical dependence Fig. 6.2.6 may be noted from a comparison.

Let us compare the results of measurements and evaluations of the radio wave's absorption ($\lambda = 2$ cm) in the clouds in the communication links satellite-satellite with the experimental data, obtained by the ground based tools for vertical communication links the earth satellite. Attenuation of centimeter radio waves in the communication link earth- satellite is small, its measurements are difficult, however there are detailed experimental investigations of the atmospheric radio emission by usage of antenna oriented in zenith [196,197]. These papers contained information concerning the total absorption of centimeter radio waves. If the absorbing medium has the temperature T, then the excess of the antenna temperature because the radio emission of medium ΔT is connected with the attenuation Y via a relationship

$$Y = 4.34\ (\Delta T)\ T^{-1}, \qquad (6.2.18)$$

where Y is in decibels, and temperature is in Kelvin. Eq. (6.2.18) is valid when absorption is small. For determination of the attenuation of centimeter waves the experimental values ΔT for the case of the stratus clouds will be applied. Values of ΔT , obtained at λ equal to 0.8; 1.6 and 3.2 cm are given in [196, 197] for the stratus clouds of the type St and Sc. For $\lambda \geq 0.8$ cm the relationship ~ λ^{-2} is valid, this allows determining the attenuation for $\lambda = 2$ cm. This simple recalculation was fulfilled from the indicated wavelengths to $\lambda = 2$ cm, it was found, that for the clouds of the type St or Sc in this period in average the attenuation is equal to $Y = 0.04$ dB, and the maximal value $Y = 0.13$ dB. Therefore, the experimental average value Y corresponds to the chosen approximated model of the stratus clouds, Eqs. (6.2.16) and (6.2.17) and, hence, to data in Fig. 6.2.5.

Let us consider the contribution of different phenomena in general attenuation of signal Y_s at the wavelength $\lambda = 2$ cm. The integral attenuation depends on the refractive attenuation X, on the absorption by oxygen Y_0 and on influence of the water vapor and the clouds Y

$$Y_s(H) = X(H) + Y_o(H) + Y(H)\,. \qquad (6.2.19)$$

Here the minimal altitudes of the radio ray H is determined from satellites position and velocity knowledge with accounting for the refraction radio waves in the atmosphere, and value X , Y_0 and Y are expressed in decibels. The refractive attenuation has been analyzed in the papers [28, 29] and described in section 2.2, where dependence $X(H)$ is given. Essentially, that the refractive attenuation is independent on the wavelength, it is determined by vertical profile of the refractivity $N(h)$ and therefore depends on the meteorological situation in the investigated regions, attenuation $X(H)$ depends on the distances between the satellites and the planetary limb L_l and L_g . In Fig. 6.2.10 curve 2 describes the theoretical dependence $X(H)$ if the near surface value $N_0 = 3.4 \cdot 10^{-4}$ and L_l = 2100 km. The altitude dependence of the oxygen's absorption is expressed by the relationship (6.2.9), and is described by curve 1 in Fig. 6.2.10. The approximated values of the oxygen's absorption for $\lambda = 2$ cm is equal to 5.5; 3.8 and 2.6 dB at the altitudes H equal to 1; 3 and 5 km, respectively. Values of the oxygen absorption Y_0 are sufficiently reliable, since the parameters $\gamma_0(0)$ and H_0 are weakly depending on the meteorological conditions. The complex attenuation because defocusing and oxygen influence $X(H)+Y_0(H)$ is indicated by curve 3 in Fig. 6.2.10.

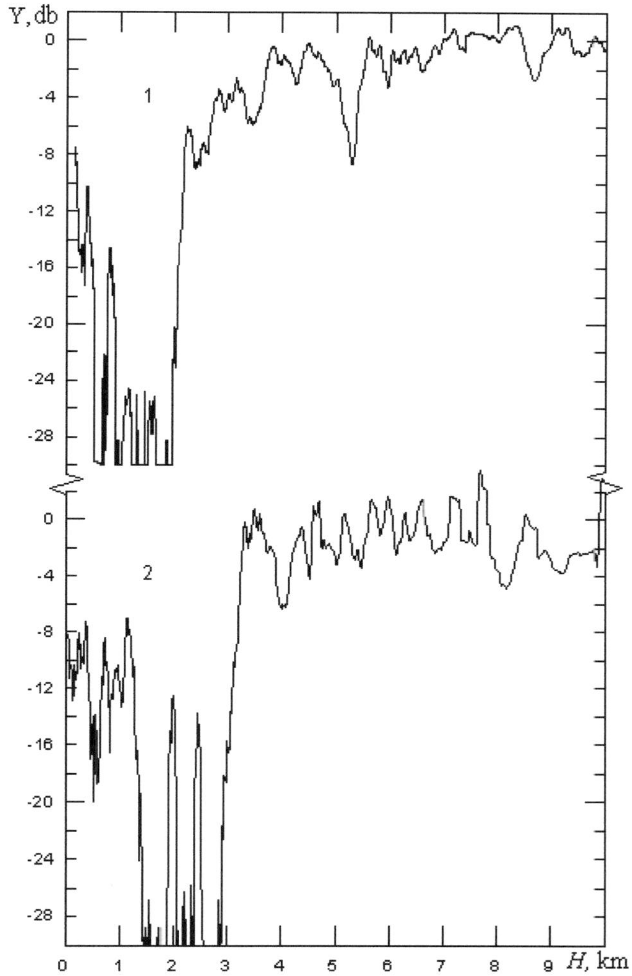

Fig.6.2.9. Examples of the intensive contribution of the stratus clouds in the radio wave absorption.

The averaged dependence of the attenuation of the decimeter radio waves $X+Y_0$, obtained from the multiple RO events, is indicated in Fig. 6.2.10 by curve 4. From comparison of dependences 3 and 4 it follows their satisfactory correspondence, distinction in data is observed only when $H<1.5$ km. The results of the experimental determination of the averaged dependence of the full attenuation of centimeter radio waves ($\lambda=2$ cm) in the communication link MIR- geostationary satellites, $Y_s(H)$ are presented by curve 5. From comparison of dependence 5 with curves 3 or 4 it follows that additional absorption of centimeter radio waves, caused by the water vapor and the clouds $Y = Y_s - (Y_0 + X)$, in average has values 6.4; 4.4 and 2.6 dB for the altitude H equal to 1; 3 and 5 km, respectively. When $H \geq 8$ km the experimental values of the integral attenuation Y_s for decimeter and of centimeter radio waves distinguishes insignificantly. dependences, shown in Fig. 6.2.10, give average values of the different components of the radio wave attenuation;

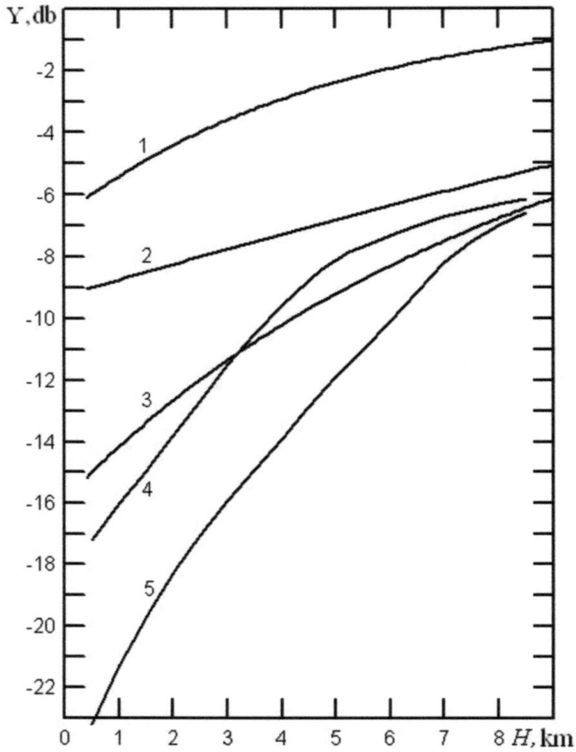

Fig.6.2.10. Different components of the attenuation of radio wave at wavelength $\lambda = 2$ cm.

during the RO measurements in regions with different meteorological conditions dependences $X(H)$ and $Y(H)$ may deviate from these averaged laws. Along the RO ray satellite-to-satellite one may determine the radio wave absorption owing to the different atmospheric gas constituents and, hence, obtain information on vertical distribution of water vapor concentration and water content of the clouds for different regions of the Earth. The absorption coefficient by water vapor γ_w and the clouds γ_c are proportional to the water vapor concentration and water content in the clouds, therefore it is possible the RO monitoring of these atmospheric parameters. From measurement data it is necessary exclude the influence of the refractive attenuation $X(H)$ and oxygen absorption $Y_0(H)$. For avoiding of these interfering factors it is necessary to apply the RO technique at two frequencies, when it is possible determination of the difference of attenuation Y_s measured at two wavelengths λ_1 and λ_2. It follows from (6.2.19)

$$Y_s(\lambda_1)-Y_s(\lambda_2)=[Y(\lambda_1)-Y(\lambda_2)]+[Y_o(\lambda_1)-Y_o(\lambda_2)] \text{ dB}, \tag{6.2.22}$$

where the first term marked by the square brackets, corresponds to the difference of the attenuations by water vapor and the clouds, and the second- to the difference of the radio wave absorption by oxygen in the ranges λ_1 and λ_2. If the conditions are fulfilled

$$Y(\lambda_1) >> Y(\lambda_2) \quad \text{and} \quad Y(\lambda_1) >> Y_0(\lambda_1)-Y_0(\lambda_2), \tag{6.2.23}$$

then, according to Eq. (6.2.22), the difference of the measured attenuation at the two wavelengths will determine the absorption of the radio waves in the clouds and water vapor. Selection of the ranges λ_1 and λ_2 for monitoring of the water vapor and the clouds is determined by dependences of γ_w, γ_0 and γ_c on the wavelength λ. With this aim in mind it is necessary to fulfill the conditions (6.2.23) and account for that the general attenuation Y_s does not should be inadmissible large. It follows from Fig. 6.2.3 that for $\lambda = 1.35$ cm in the moderate latitudes at the altitudes $H = 10$-15 km in summer time the absorption $Y_w \approx 6$ dB, and in winter time $Y_w \approx 3$ dB; at the altitude $H = 5$ km in summer time $Y_w \approx 17$ dB, and in winter time $Y_w \approx 8$ dB. In the range $\lambda = 1.35$ cm, when $H < 3$ km, the absorption become inadmissible large even in the winter time conditions. Therefore the optimal range radio waves for RO monitoring of the water vapor and the clouds is located in sharp intervals $\lambda_1 \approx 1.5$-2.1 cm and $\lambda_2 \approx 5$-10 cm. The experimental determination $Y(H)$ with by the dual frequency technique allows with usage of the Abel transformation finding the altitude profile of the complete absorption coefficient of the radio wave by water vapor and clouds $\gamma_w + \gamma_c$ without usage of any other model of their vertical distributions. One can conclude, that the described method gives possibility to determine the absorption of centimeter radio waves at the different altitudes of the radio ray perigee, therefore it is possible monitoring of the humidity and the clouds, i.e. determination of their water content, the upper and lower boundaries, and also finding of the water vapor content. For solution of this problem it is necessary to apply a method which allows reliable excluding the interfering influence of the defocusing, decreasing the influence of the amplitude fluctuations and absorption by the oxygen. The refractive attenuation significantly depends on the meteorological situation in the regions of RO sounding, and is independent on the wavelength; therefore exception of its influence is possible only by usage of dual frequency measurements. For elaboration of this method of monitoring the water vapor and the clouds it is necessary to have defined more accurately values of the absorption coefficients of water vapor $\gamma_w(\lambda)$ and the clouds $\gamma_c(\lambda)$ for the centimeter range. Note the basic technical difficulty in the RO monitoring of the clouds and water vapor. During radio occultation the lower troposphere is observed in condition of the large general attenuation of radio waves. With accounting for the amplitude fluctuations the attenuation relative to the free radio wave propagation at $\lambda = 2$ cm achieved 25-28 and 18-22 dB at the altitudes H equal to 0.5 and 2 km, respectively , therefore it is necessary to use the radio links with large energetic potential.

A new possibility of determination of the radio wave's absorption in the case of one frequency radio occultation has been shown in [55]. This method is based on a connection of variations of the phase and refractive attenuation. It was indicated in chapter 3, 4, that the altitude dependence of the phase allows determining then refractive attenuation $X_p(H)$ since the second derivative phase with respect to time gives dependence $X(H)$. The experimental dependence of the signal intensity on time, after its normalization, gives dependence of the combined absorption and refractive attenuation $Y+X$. Registration of the phase variations allows determining the refractive attenuation X_p. Determination of the sum $Y+X$ from the amplitude data and X_p from the phase measurements allows excluding the interfering influence of the refractive attenuation and determining the integral radio wave absorption $Y(H)$. If applied to the decimeter waves, this method may determine the radio wave absorption by oxygen. Application to the centimeter waves allows more detailed studies of radio wave absorption by the clouds having different structure and by the water vapor.

7. References

1. Fjeldbo, G., and v.R. Eshleman (1965), The bistatic radar-occultation method for the study of planetary atmospheres, *J. Geophys. Res., 70*(13), 3217.
2. Kunitsyn v.E., Tereschenko E.D. (2003) Ionospheric tomography. Springer Verlag.
3. Fjeldbo, G., W. Fjeldbo, and R. Eshleman (1966), Models for the atmosphere of Mars based the Mariner-4 occultation experiments, *J. Geophys. Res., 71*(9), 2307.
4. Kliore, A., G. Fjeldbo, B. Seidel, and S. Rasool (1969), Mariner 6 and 7: radio occultation measurements of the atmosphere of Mars, *Science, 166*(3911), 1393.
5. Kolosov, M.A., O.I. Yakovlev, Yu.M. Kruglov et al. (1972), Radio occultation of the atmosphere of Mars by use of spacecraft Mars-2, *Radiotehnica i electronica, 17*(12), 2483, (in Russian).
6. Kliore, A., G. Fjeldbo, B. Seidel (1970), First results of Mariner-6 radio occultation measurements of the lower atmosphere of Mars, *Radiosci.*, 5(2), 373.
7. Fjeldbo, G., and v.R. Eshleman (1969), Atmosphere of Venus as studied with the Mariner-5 dual radio-frequency occultation experiment, *Radio Sci., 4*(10), 879.
8. Fjeldbo, G., B. Seidel, D. Sweetnam, and T. Howard (1975), The Mariner-10 radio occultation measurements of the ionosphere of Venus, *J. Atmos. Sci., 32*(6), 1232.
9. Kolosov, M.A., O.I. Yakovlev, B.P. Trusov et al. (1976), Radio occultation investigation of the atmosphere of Venus by use of satellites Venera-9 and Venera-10, *Radio Eng. and Electron. Phys., 21*(8), 1585.
10. Kolosov, M.A., O.I. Yakovlev, A.I. Efimov et al. (1979), Radio occultation of the atmosphere and bistatic radiolocation of the Venus surface using the Venera 9 and 10 satellites, *Radio Sci., 14*(1), 163.
11. Kolosov, M.A., O.I. Yakovlev, A.I. Efimov et al. (1980), Investigation of the Venus atmosphere by radio occultation method using Venera-9 and 10 satellites, *Acta Astron., 7*, 219.
12. Pavelyev, A.G., and A.I. Kucherjavenkov (1994), Bistatic sounding of planetary surfaces, *Itogi nauki i techniki. Radiotechnica, 44*, 81–175, vINITI Ed., (in Russian).

13. Ivanov, G.S., M.A. Kolosov, N.A. Savich et al. (1979), Daytime ionosphere of Venus as studied with Venera 9 and 10 dual-frequency experiments, *Icarus, 39*(2), 209.
14. Kliore, A.I. (1985), Recent results on the Venus of the atmosphere from Pioner-Venus radio occultations, *Adv. Space Res., 5*(9), 41.
15. Yakovlev, O.I., S.S . Matyugov, and v.N. Gubenko (1991), Venera-15 and 16 radio occultation investigations of the middle atmosphere in the polar and near-polar regions of Venus, *Icarus, 94*, 493.
16. Yakovlev, O.I. (2003), *Space Radio Science*, 306 pp., Taylor and Francis, London.
17. Pavelyev, A.G., S.G. Rubashkin, A.I. Kucherjavenkov, R.R. Salimzyanov, and D.A. Pavelyev (1993), Results of measurement of the refractive angle of radio waves in the atmosphere of Venus from bistatic radar data, *J. Commun. Techn. Electron., 38*(15), 43.
18. Kliore, A. (1969), Some remarks on meteorological measurements with occultation satellites, *Space Res, 9*, 590.
19. Lusignan, B., G . Modrell, A . Morrison et al. (1969), Sensing the earth's atmosphere with occultation satellites, *Proc. IEEE, 57*(4), 458.
20. Ungar, S.G., and B.B. Lusignan (1973), A two-satellite microwave occultation system for determining pressure altitude references, *J. Appl . Meteorol., 12*(2), 396.
21. Kalashnikov, I., and O.I. Yakovlev (1978), On a possibility of investigation of the earth's atmosphere by radio occultation method, *Cosmic Res., 16*(6), 943.
22. Kalashnikov, I., S . Matyugov, A. Pavelyev, and O. Yakovlev (1986), Analysis of RO method for the earth's atmosphere study, In the book *Electromagnetic waves in the atmosphere and space*, Nauka, Moscow, 208–218 (in Russian).
23. Gurvich, A.S., and T.G. Krasilnikova (1987), Application of the navigational satellites for radio occultation of the earth's atmosphere, *Cosmic Res.*, 6, 89.
24. Yunck, T.P., G. F. Lindal, and C. H. Liu (1988), in *IEEE Position, Location and Navigation Symp., Orlando, FL., Nov.29–Dec. 2*, (IEEE, Piscataway, NJ, 1988), 251.
25. Liu, A.S. (1978), On determination and investigation of the terrestrial ionospheric refractive indices using Geos-3/ATS-6 satellite-to-satellite tracking data, *Radio Sci, 13*(4), 709.
26. Rangaswamy, S. (1976), Recovery of atmospheric parameters from the Apollo/Soyuz-ATS-F radio ocultation data, *Geophys. Res. Letters, 3*(8), 483.
27. Yakovlev, O.I., v.A. Grishmanovskiy, S.D. Eliseev et al. (1990), Radio occultation of the earth's atmosphere by use of two satellites, *Sov.Phys.Dokl., 315*(1), 101.
28. Yakovlev, O.I., I.A. Vilkov, v.A. Grishmanovskiy et al. (1992), Refractive attenuation in the radio occultation remote sensing the earth's atmosphere in the communication link satellite-to-satellite, *J. Commun. Techn. Electron., 37*(1), 42.
29. Yakovlev, O.I., S.S . Matyugov, and I.A. Vilkov (1995), Attenuation and scintillation of radio waves in the atmosphere from data of radio occultation experiments in the satellite-to-satellite link, *Radio Sci., 30*(3), 591.
30. Pavelyev, A.G., A.V. Volkov, A.I. Zakharov, S.A. Krytikh, and A.I. Kucherjavenkov (1996), Bistatic radar as a tool for the earth's observation using small satellites, *Acta Astron., 39*(9–12), 721.
31. Pavelyev, A.G., A.I. Zakharov, A.I. Kucherjavenkov, A.I. Sidorenko, I.L. Kucherjavenkova, and D.A. Pavelyev (1997), The features of propagation of radio wave reflected from terrestrial surface at small elevation angles on radio telecommunication link low orbital satellite-GEO, *J. Commun. Techn. Electron., 42*(1), 51.

32. Ware, R., M. Exner, D. Feng, M. Gorbunov, K. Hardy, B. Herman, Y. Kuo, T . Meehan, W . Melbourne, C. Rocken, W. Schreiner, S. Sokolovskiy, F. Solhelim, X. Zou, R. Anthes, S. Businger, and K. Trenberth, (1996), GPS sounding of the atmosphere from low earth orbit: preliminary results, *Bull. Am . Met. Soc., 77*(1), 19.
33. Leroy, S.S., (1997), Measurement of geopotential heights by GPS radio occultation, *J. Geophys. Res., 102*, 6971.
34. Rius, A., G. Ruffini, and A. Romeo (1998), Analysis of ionospheric electron-density distribution from GPS/MET occultations, *IEEE Trans. Geosci. Remote Sens., 36*(2),383.
35. Schreiner, W.S., S.S. Sokolovskiy, C. Rocken, and R.H. Ware (1999), Analysis and validation of GPS/MET radio occultation data in the ionosphere, *Radio Sci., 34*(4), 949.
36. Kursinski, E.R., G.A. Hajj, W.L. Beritger et al. (1999), Initial results of radio occultation observations of the earth's atmosphere using GPS, *Science, 271*, 1107.
37. Kursinski, E.R, G.A. Hajj, K.R. Hardy et al. (1997), Observing the earth's atmosphere with radio occultation measurements using GPS, *J. Geophys. Res., 102*(D19), 23492.
38. Rocken, C., R. Anthes, M. Exner et al. (1997), Analysis and validation of GPS/MET data in the neutral atmosphere, *J. Geophys. Res., 102*, 29849.
39. Palmer, P.I., J.J. Barnett, J.R. Eyre, and S.B. Healy (2000), A nonlinear optimal, estimation inverse method for radio occultation measurements of temperature, humidity, and surface pressure, *J. Geophys. Res., 105*(D13), 17513.
40. Anthes, R.A., C. Rocken, and Y. Kuo (2000), Applications of COSMIC to Meteorology and Climate, *Terrestrial Atmospheric and Oceanic Science, 11*(1), 115.
41. Leroy, S.S. and G.R. North (2000), The application of COSMIC data to global change research, *Terrestrial Atmospheric and Oceanic Science, 11*(1), 235.
42. Mostert, S. and J.A. Koekemoer (1997), The science and engineering payloads and experiments on SUNSAT, *Acta Astronautica, 41*(4 -10), 401.
43. Kursinski, E.R. and G. A. Hajj (2001), A comparison of water vapor derived from GPS occultations and global weather analyses, *J. Geophys. Res., 106*(D1), 1113.
44. Escudero, A., A.C. Schlesier, A. Rius, A. Flores, F. Rubek, G.B. Larsen, S. Syndergaard, and P. Hoeg (2001), Ionospheric tomography using Ørsted GPS measurements - Preliminary results, *Physics and Chemistry of Earth A, 26*(3),173.
45. Hajj, G.A., C.O. Ao, B.A. Iijima, D. Kuang, E.R. Kursinski, A.J . Mannucci, T.K . Meehan, L.J. Romans, M. de la Torre Juárez, and T.P. Yunck (2004), CHAMP and SAC-C atmospheric occultation results and intercomparisons, *J. Geophys. Res.*, 109(D06109), doi: 10.129/2003JD003909.
46. Reigber, G., H. Luhr, and P. Schwintzer (2003), *First CHAMP mission results*, 562pp., Springer.
47. Wickert, J., T. Schmidt, G. Beyerle, G . Michalak, R. König, S. Heise, and Ch. Reigber (2006), GPS radio occultation with CHAMP and GRACE: Recent results, in *Atmosphere and Climate Studies by Occultation Methods*, edited by U. Foelsche, G. Kirchengast, and A.K. Steiner, 3–16, Springer Verlag.
48. Hajj, G.A., E.R. Kursinski, L.J. Romans, W.I. Bertiger, and S.S. Leroy (2002), Technical Description of Atmospheric Sounding by GPS occultation, *J. Atmos. Solar-Terr. Phys.*, *64*, 451.

49. Anthes, R., P. Bernhardt, Y. Chen, L. Cucurull, K. Dymond, D. Ector, S. Healy, S.-P. Ho, D. Hunt, Y.-H. Kuo, H. Liu, K . Manning, C . McCormick, T . Meehan, W. Randel, C. Rocken, W. Schreiner, S. Sokolovskiy, S. Syndergaard, D. Thompson, K. Trenberth, T.-K. Wee, N. Yen, and Z. Zhang (2008), *The COSMIC/FORMOSAT-3 mission: Early results. Bull. Am . Met. Soc. (March)*, 1.
50. B. Parkinson and J. Spilker, eds., Global Positioning System: Theory and Application, American Institute of Aeronautics and Astronautics,Washington, DC, 1996.
51. W. G . Melbourne, E. S. Davis, C. B. Duncan, G. A. Hajj, K. R. Hardy, E. R. Kursinski, T. K . Meehan, L. E. Young, and T. P. Yunck, The Application of Spaceborne GPS to Atmospheric Limb Sounding and Global Change Monitoring, JPL Publication 94-18, Jet PropulsionLaboratory, Pasadena, California, April 1994.
52. Melbourne, W.G. (2004), *Radio Occultations Using Earth Satellites: A Wave Theory Treatment*, Monograph 6 Deep space communications and navigation series, Jet Propulsion Laboratory of California Institute of Technology.
53. Pavelyev, A.G., and A. I. Kucherjavenkov (1978), Refractive attenuation in the planetary atmospheres, *Radio Eng. Electron. Phys.*, *23*(7), 13.
54. Eshleman, v.R., D.O . Muhleman, P.D. Nicholson, and P.G. Stesses (1980), Comment on absorbing regions in the atmosphere of Venus as measured by radio occultation, *Icarus*, *44*, 793.
55. Liou, Y.A., and A.G. Pavelyev (2006), Simultaneous observations of radio wave phase and intensity variations for locating the plasma layers in the ionosphere, *Geophys. Res. Lett.*, *33*, L 23102.
56. Phinnely, R.A., and D.L. Anderson (1968), On the radio occultation method for studying planetary atmospheres, *J. Geophys. Res.*, *73*(5), 1819.
57. Hays, P.B., and R.G. Roble (1968), Atmospheric properties from inversion of planetary occultation data, *Planet. Space Sci.*, *16*(9), 1197.
58. Fjeldbo, G., and v.R. Eshleman (1968), The atmosphere of Mars analyzed by integral inversion of the Mariner-4 occultation data, *Planet Space Sci.*, *16*(8), 1035.
59. Yakovlev, O.I., v.P. Yakubov, v.P. Uriadov, and A.G. Pavelyev (2009), *Propagation of radio waves*, 406 pp., URSS Ed., Moscow, (in Russian).
60. Smith, E.K., and S. Weintraub (1953), The constants in the equation for atmospheric refractive index at radio frequencies, *Proc. IRE.*, *41*, 1035.
61. Bean, B.R., and E.J. Dutton (1966), 431 pp., *Radio Meteorology*, National bureau of standards, Washington.
62. Kolosov, M.A., and A.V. Shabelnikov (1976), *Refraction of electromagnetic waves*, Soviet Radio, Moscow, (in Russian).
63. Bilitza, D. (2001), International Reference Ionosphere, *Radio Sci.*, *36*(2).
64. Bilitza, D., and B.W. Reinisch (2008), International reference ionosphere 2007, *Adv. Space. Res.*, *42*, 599.
65. Pavelyev, A.G., K. Igarashi, C. Reigber, K. Hocke, J. Wickert, G. Beyerle, S. Matyugov, A. Kucherjavenkov, D. Pavelyev, and O.I. Yakovlev (2002), First application of the radio holographic method to wave observations in the upper atmosphere, *Radio Sci.*, *37*(3), 15–1.

66. Liou, Y.-A., A.G. Pavelyev, C.-Y. Huang, K. Igarashi, and K. Hocke (2002), Simultaneous observation of vertical gradients of refractivity in the atmosphere and electron density in the lower ionosphere by radio occultation amplitude method, *Geophys. Res. Lett.*, *29*(19), 43–1–43–4, doi:10.1029/2002GL015155.
67. Hocke, K. (1997), Inversion of GPS meteorology data, *Ann. Geophys.*, *15*, 443.
68. Zverev, v. A. (1975), *Radio-optics*, Soviet Radio, Moscow, (in Russian).
69. Pavelyev, A. (1998), On possibility of radio holographic investigation on communication link satellite-to-satellite, *J. Commun. Techn. Electron.*, *43*(8), 126.
70. Hocke, K., A.G. Pavelyev, O.I. Yakovlev et al. (1999), Radio occultation data analysis by the radioholographic method, *J. Atmosp. Sol. –Terr. Phys.*, *61*(15), 1169.
71. Beyerle, G., and K. Hocke (2001), Observation and simulation of direct and reflected GPS signals in radio occultation experiments, *Geophys. Res. Lett.*, *28*(9), 1895.
72. Beyerle, G., K. Hocke, J. Wickert et al. (2002), GPS radio occultation with *CHAMP*: A radio holographic analysis of GPS signal propagation in the troposphere and surface reflections, *J. Geophys. Res.*, *107*(D24), doi: 10.1029/2001JD001402, 27–1.
73. Igarashi, K., A. Pavelyev, K. Hocke, D. Pavelyev, and J. Wickert (2001), Observation of wave structures in the upper atmosphere by means of radio holographic analysis of the radio occultation data, *Adv. Space Res.*, *27*(6–7), 1321.
74. Pavelyev, A.G., Y.A. Liou, and J. Wickert (2004), Diffractive vector and scalar integrals for bistatic radio- holographic remote sensing, *Radio Sci.*, *39*(4), RS4011, 1–16.
75. Jensen, A.S., M. S. Lohmann, H.-H. Benzon, and A. S. Nielsen (2003), Full Spectrum Inversion of Radio Occultation Signals, *Radio Sci. 38*(3), doi: 10.1029/2002RS002763.
76. Jensen, A.S., M.S. Lohmann, A.S. Nielsen, and H.-H. Benzon (2004), Geometrical Optics Phase Matching of Radio Occultation Signals, *Radio Sci.*, *39*, RS3009, doi: 10.1029/2003RS002899, 1–8.
77. Lohman, M.S., A.S. Jensen, H.H. Benson, and A.S. Nielsen (2003), *Radio Occultation retrieval of Atmospheric Absorption based on FSI Report 03–22*, Danish Meteorological Institute, Copenhagen.
78. Gorbunov, M.E., and G. Kirchengast (2007), Fluctuations of radio occultation signals in X/K band in the presence of anisotropic turbulence and differential transmission retrieval performance, *Radio Sci.*, *42*(5), doi: 10.1029/2000RS002592, 9–1.
79. Gorbunov, M.E. and K.B. Lauritsen (2004), Analysis of wave fields by Fourier integral operators and its application for radio occultations, *Radio Sci.*, *39*(4), RS4010, doi: 10.1029/2003RS002971.
80. Gorbunov, M. E., and A.S. Gurvich (1998), Algorithms of inversion of Microlab-1 satellite data including effects of multipath propagation, *Internat. J. Rem. Sens.*, *19*(12), 2283.
81. Gorbunov, M.E. (2002), Canonical transform method for processing GPS radio occultation data in lower troposphere, *Radio Sci.*, *37*(5), doi: 10.1029/2000RS002592, 9-1.
82. Pavelyev, A.G., Y.A. Liou, J. Wickert, T. Schmidt, A.A. Pavelyev, and S.F. Liu (2007), Effects of the ionosphere and solar activity on radio occultation signals: Application to CHAllenging Minisatellite Payload satellite observations, *J. Geophys. Res.*, *112*(A06326), 1–14.
83. Liou, Y.A., A.G. Pavelyev, S.F. Liu, A.A. Pavelyev, N. Yen, C.Y. Huang, and C.J. Fong (2007), FORMOSAT-3 GPS radio occultation mission: preliminary results, *IEEE Trans. Geosci. Rem. Sens.*, *45*(10), 3813.

84. Gorbunov, M.E., A.S. Gurvich, and A.V. Shmakov (2002), Back-propagation and radio-holographic methods for investigation of sporadic ionospheric E-layers from Microlab-1 data, *Int. J. Rem. Sens., 23*(4), 675.
85. Sokolovskiy, S.V., W. Schreiner, C. Rocken, and D. Hunt (2002), Detection of high-altitude ionospheric irregularities with GPS/MET, *Geophys Res Lett, 29*(3), 1033.
86. Martini, E., A. Freni, L. Facheris, and F. Cuccoli (2006), Impact of tropospheric scintillation in the Ku/K bands on the communications between two LEO satellites in a radio occultation geometry, *IEEE Trans. Geosci. Rem. Sens., 44*(8), 2063.
87. Kirchengast, G., and P. Hoeg (2004), *The ACE+ mission: Atmosphere and climate explorer based on GNSS-LEO and LEO-LEO radio occultation, in Occultations for Probing Atmosphere and Climate,* edited by A. K. S. G. Kirchengast and U. Foelsche, pp. 201–220, Springer, New York.
88. Gorbunov, M.E. and G. Kirchengast (2005), Processing X/K Band Radio Occultation Data in Presence of Turbulence, *Radio Sci., 40*(6), RS6001, doi: 10.1029/2005RS003263.
89. Pavelyev, A.G., Y.A. Liou, J. Wickert, A.A. Pavelyev, T. Schmidt, K. Igarashi, and S.S . Matyugov (2008), Location of layered structures in the ionosphere and atmosphere by use of GPS occultation data, *Adv. Space Res., 42*, 224.
90. Pavelyev, A.G., J. Wickert, and Y. Liou (2008), Localization of plasma layers in the ionosphere based on observing variations in the phase and amplitude of radiowaves along the satellite-to-satellite path, *Radiophys. Quant. Electron., 51*(1), 1.
91. Pavelyev, A.G., S.S . Matyugov, and O.I. Yakovlev (2008), Global Satellite Monitoring of the Atmosphere and Ionosphere, *J. Commun. Techn. Electron.,* Pleiades Publishing, *53*(9), 1021.
92. Vorob'ev, v.V. and T.G. Krasilnikova (1994), Estimation of accuracy of the atmosphere refractive index recovery from Doppler shift measurements at frequencies used in the NAVSTAR system, *Izvestiya Russian Academy of Sciences, Physics of the Atmosphere and Ocean*, Engl. Transl., *29*(7), 602.
93. Gorbunov, M.E. (2005), *Diffraction theory of radio occultation sounding of the earth's atmosphere*, 292 pp., Dissertation . Moscow, (in Russian).
94. Sokolovskiy, S., C. Rocken, D. Hunt, W. Schreiner, and J. Johnson (2006), GPS profiling of the lower troposphere from space: Inversion and demodulation of the open-loop radio occultation signals, *Geophys. Res. Lett., 33*(L14816), doi:10.1029/2006GL026112, 1–4.
95. Spilker, J.J., (1980), GPS signal structure and performance characteristics, in *Global Positioning System*, edited by P.M. Janiczek, 29–54, Inst. of Navig., Washington, D.C.
96. Gobiet, A. and G. Kirchengast (2004), Advancements of Global Navigation Satellite System radio occultation retrieval in the upper stratosphere for optimal climate monitoring utility, *J. Geophys. Res., 109*(D24110), doi:10.1029/2004JD005117, 1–11.
97. Pavelyev, A.G., A.V. Volkov, A.I. Zakharov, and A.I. Kucherjavenkov (1996), Bistatic radar as a tool for earth investigation using small satellite, *Acta Astron., 39*, 721.
98. Igarashi, K., A.G. Pavelyev, K. Hocke, D.A. Pavelyev, A.I. Kucherjavenkov, S.S . Matyugov, A. I. Zakharov, and O.I. Yakovlev (2000), Radio holographic principle for observing natural processes in the atmosphere and retrieving meteorological parameters from radio occultation data, *Earth Planets Space, 52*, 893.
99. Sokolovskiy, S. (2000), Inversion of radio occultation amplitude data, *Radio Sci., 35*(1), 97.

100. Steiner, A.K. and G. Kirchengast (2000), GW spectra from GPS/MET occultation observations, *J. Atmos. Ocean. Tech., 17*, 495.
101. Tsuda, T. and K. Hocke (2002), vertical wave number spectrum of temperature fluctuations in the stratosphere using GPS occultation data, *J . Met.. Soc. Japan., 80*(4B), 1.
102. Luers, J.K., and Eskridge R.E. (1995), Temperature corrections for the vIZ and vaisala radiosondes, *J. Appl . Meteorol., 34*, 1241.
103. Smith, W.L., W.M. Woolf, C.M. Hayden et al. (1979), The TIROS-N operational sounder, *Bull. Am . Met.. Soc., 60*, 1177.
104. Steiner, A., A. Loscher, and G. Kirchengast (2006), Error characteristics of refractivity profiles retrieved from CHAMP radio occultation data, in the book *Atmosphere and Climate Studies*, Kirchengast G., Foelsche U.; Steiner A. (Eds.), Springer-Verlag.
105. Nishida, M., T. Tsuda, C. Rocken et al. (2000), verification of temperature profiles obtained by GPS occultation technique with radiosonde and lidar measurements in tropical and subtropical regions, *J . Met. Soc. Japan, 78*, 691.
106. Bizzarri, B., I. Bordi, A. Dell'Aquila et al. (2006), Analysis of seasonal and daily mid – latitude tropopause pressure using GPS radio occultation data and NCEP – NCAR reanalyses, In the book *Atmosphere and Climate Studies*, Kirchengast G., Foelsche U.; Steiner A. (Eds.) Springer-Verlag.
107. Schmidt, T., J. Wickert, G. Beyerle, and C. Reigber (2004), Tropical tropopause parameters derived from GPS radio occultation measurements with CHAMP, *J. Geophys. Res., 109*(D13105), doi: 10.1029/2004 JD004546.
108. Kalnay, E., M. Kanamitsu, R. Kistler et al. (1996), The NCEP/NCAR 40-year reanalysis project, *Bull. Am . Met. Soc., 77*, 437.
109. Nagpal, O.P. (1979), The sources of atmospheric GWs, *Contemp. Phys., 20*, 593.
110. Friits, D.C., and M.J. Alexander (2003), Gravity wave dynamics and effects in the middle atmosphere, *Rev. Geophys., 41*, 3-1–3-64.
111. Ebel, A. (1984), Contribution of GWs to the momentum, heat and turbulent energy budget of the upper mesosphere and lower thermosphere, *J. Atmos. Terr. Phys., 46*(9), 727.
112. Pruse, J.M., P.K. Smolarkiewicz, and R.R. Garcia (1996), Propagation and breaking at high altitudes of GWs excited by tropospheric forcing, *J. Atmos. Sci., 53*, 2186.
113. Murayama, Y., T. Tsuda, S. Kato, and S. Fukao (1994), Seasonal variation of gravity wave activity in the lower stratosphere observed with the MU radar, *J. Geophys. Res., 99*, 23057.
114. Pfister, L., K.R. Chan, T.B. Bui et al. (1993), Gravity waves generated by a tropical cyclone during the STEP tropical field program, *J. Geophys. Res., 98*, 8611.
115. Hamilton, K. (1991), Climatological statistics of stratospheric inertia-gravity waves deduced from historical rocketsonde wind and temperature data, *J. Geophys. Res., 96*, 20831.
116. Eckermann, S.D., I. Hirota, and W.A. Hocking (1995), GW and equatorial wave morphology of the stratosphere derived from long-term rocket soundings, *Q.J.R . Meteorol. Soc., 121*, 149.
117. Allen, S.J., and R.A. Vincent (1995), Gravity wave activity in the lower atmosphere: Seasonal and latitudinal variations, *J. Geophys. Res., 100*, 1327.
118. Sica, R.J., and A.T. Russell (1999), Measurements of the effects of gravity waves in the middle atmosphere using parametric models of density fluctuations. Part I: vertical Wavenumber and Temporal Spectra, *J. Atmos. Sci., 56*, 1308.

119. Eckermann, S.D., and P. Preusse (1999), Global measurements of stratospheric mountain waves from space, *Science, 286*, 1534.
120. Wu, D.L., and J.W. Waters (1996), Satellite observations of atmospheric variances: A possible indication of gravity waves, *Geophys. Res. Lett., 23*, 3631.
121. Pavelyev, A.G., J. Wickert, Y.A. Liou et al. (2006), Analysis of atmospheric and ionospheric wave structures using the CHAMP and GPS/MET radio occultation database, In the book *Atmosphere and Climate Studies*, Kirchengast G., Foelsche U.; Steiner A. (Eds.) Springer- Verlag.
122. Tsuda, T., M. Nishida, C. Rocken, and R.H. Ware (2000), A global morphology of GW activity in the stratosphere revealed by the GPS occultation data (GPS/MET), *J. Geophys. Res., 105*, 7257.
123. Liou, Y.A., A.G. Pavelyev, K. Igarashi K. et al. (2003), Analytic method for observation of the gravity waves using radio occultation data, *Geophys. Res. Lett., 30*(20), ASC 1-1 - 1-5.
124. Liou, Y.A., A.G. Pavelyev, and J. Wickert (2005), Observation of gravity waves from GPS/MET radio occultation data, *J. Atmos. Sol. Terr. Phys., 67*, 219.
125. Liou, Y.A., A.G. Pavelyev, J. Wickert, et al. (2006), Application of GPS radio occultation method for observation of the internal waves in the atmosphere, *J. Geophys. Res., 111*(D06104), doi:10.1029/2005JD005823.
126. Rabiner, L, and B. Gold (1978), *Theory and application of digital signal processing*, Prentice-Hall, Englewood Cliffs.
127. Afanasyev, Ya.D. and W.R. Peltier (2001), Numerical Simulation of Internal Gravity Wave Breaking in the Middle Atmosphere: The Influence of Dispersion and Three-Dimensionalization, *J. Atmos. Sci., 58*(2), 132.
128. Kislyakov, A.G., and K.S. Stankevich (1967), Absorption of radio waves in the atmosphere, *Izv Vyssh Uchebn Zaved. Radiophyz, 10*(9–10), 1244, (in Russian).
129. Marouf, E.A., G.L. Tyler, and Rosen P.A. (1986), Profiling Saturn rings by radio occultation, *Icarus, 68*, 120.
130. Mortensen, M.D., and P. Hoeg (1998), Inversion of GPS occultation measurements using Fresnel diffraction theory, *Geophys. Res. Lett., 25*(14), 2446.
131. Gorbunov, M.E., and L. Bengtson (1996), *Advanced algorithms of inversion of GPS/MET satellite data and their application to reconstruction of temperature and humidity. Tech. Rep Report No. 211*, Max Planck Institute for Meteorology, Hamburg.
132. Hinson, D.P., F.M. Flasar, A. Schinder, J.D. Twicken, and R.G. Herrera (1997), Jupiter's Ionosphere: Results from the first Galileo radio occultation experiment, *Geophys. Res. Lett., 24*(18), 2107.
133. Lindal, G.F., J.R. Lyons, D.N. Sweetnam, v.R. Eshleman, D.P. Hinson, and G.L. Tyler (1987), The atmosphere of Uranus: Results of radio occultation measurements with Voyager 2, *J. Geophys. Res., 92*(A13), 14987.
134. Sokolovskiy, S.V. (2001), Modeling and inverting radio occultation signals in the moist troposphere, *Radio Sci., 36*(3), 441.
135. Igarashi, K., A. Pavelyev, J. Wickert, K. Hocke, and D. Pavelyev (2002), Application of radio holographic method for observation of altitude variations of the electron in the mesosphere using GPS/MET radio occultation data, *J. Atmos. Sol. Terr. Phys., 64*(8–11), 959.

136. Pavelyev A.G., Y.A. Liou, J.Wickert, A.L. Gavrik, and C.C. Lee (2009), Eikonal acceleration technique for studying of the Earth and planetary atmospheres by radio occultation method *Geophys. Res. Lett.* 36(21), L21807, 1-5.
137. Pavelyev A.G., Y.-A. Liou, J. Wickert, T.Schmidt, A.A. Pavelyev, and S.S. Matyugov (2010), Phase acceleration: a new important parameter in GPS occultation technology. GPS Solutions, 14(1), 3-14.
138. Pavelyev A.G., Y.A. Liou, J. Wickert, v.N. Gubenko, A.A. Pavelyev, and S.S. Matyugov (2009) New Applications and Advances of the GPS Radio Occultation Technology as Recovered by Analysis of the FORMOSAT-3/COSMIC and CHAMP Data-Base. Andrea Steiner · Barbara Pirscher ·Ulrich Foelsche · Gottfried Kirchengast Editors. New Horizons in Occultation Research. Studies in Atmosphere and Climate. Springer-Verlag Berlin Heidelberg.p. 165-178.
139. Ishimaru, A. (1978), Wave propagation and scattering in random media, 2, Academic press, N.Y.
140. Woo, R., A. Ishimaru, and F. Yang (1980), Radio scintillatons during occultations by turbulent planetary atmospheres, *Radio Sci.*, *15*(3), 695.
141. Woo, R., L. Armstrong, and A. Kliore (1982), Small - scale turbulence in the atmosphere of Venus, *Icarus*, *52*(2), 335.
142. Vilkov, I.A., S.S . Matyugov, and O.I. Yakovlev (1993), Amplitude fluctuations during radio occultation investigation of the Earth's atmosphere along a satellite-to-satellite path, *J. Commun. Techn. Electron.*, *38*(5), 24.
143. Anufriev, v.A., and O.I. Yakovlev (2002), Amplitude and phase fluctuations of decimeter radio waves raying the atmosphere via satellite-to-satellite paths, *Radiophys. Quant. Electron.*, *45*(7), 501.
144. Yakovlev, O.I., S.S . Matyugov, I.A. Vilkov et al. (1996), Radio-wave phase and frequency fluctuations as observed in radio-eclipse experiments along a satellite-satellite link, *J. Commun. Techn. Electron.*, *41*(9), 993.
145. Kan, v., S.S . Matyugov, and O.I. Yakovlev (2002), The structure of stratospheric irregularities according to radio-occultation data obtained using satellite-to-satellite paths, *Radiophys. Quant. Electron.*, *45*(8), 595.
146. Gurvich, A.S. (1989), Scintillations spectra observed during stellar occultation by the earth's atmosphere, *Atmos. Optics*, 2(3), 239.
147. Alexandrov, A.P., G.M. Grechko, A.S. Gurvich et al. (1990), Spectra of temperature variations in the stratosphere from observations of stars scintillations from space, *Izvestiya Russian Academy of Sciences, Physics of the Atmosphere and Ocean*, *26*(1), 5, (in Russian).
148. Smith, S.A., D.C. Fritts, and T.E. Vanzandt (1987), Evidence for a saturated spectrum of atmospheric gravity waves, *J. Atmos. Sci.*, *44*(10), 1404.
149. Fritts, D., T. Tsuda, T. Sato et al. (1988), Observational evidence of saturated gravity wave spectrum, *J. Atmos. Sci.*, *45*(12), 1741.
150. Gurvich, A.S., v. Kan, and O.V. Fedorova (2000), Radio occultation investigation of the atmosphere by use of satellites GPS–MICROLAB: The phase fluctuations, *Izvestiya Russian Academy of Sciences, Physics of the Atmosphere and Ocean*, *36*(3), 330, (in Russian).

151. Yakovlev, O.I., v.A. Anufriev, S.S . Matyugov, and A.G. Pavelyev (2001), Atmospheric fluctuations of phase of decimeter radio waves in occultation experiments in communication links satellite-satellite, *J. Commun. Techn. Electron.*, *46*(12), 1439.
152. Matyugov, S.S., O.I. Yakovlev, and I.A. Vilkov (1994), Absorption of decimeter and centimeter radio waves during occultation investigation of the atmosphere in communication links satellite-satellite, *J. Commun. Techn. Electron.*, *39*(8–9), 1251.
153. Gurvich, A.S., v. Kan, S.S. Savchenko et al. (2001), Investigation of turbulence and internalwaves in the stratosphere from observations of scintillations of stars from space, *Izvestiya Russian Academy of Sciences, Physics of the Atmosphere and Ocean*, *37*(4), 487 and (2003), *39*(3), 335, (in Russian).
154. Gurvich, A.S. (1984), Fluctuations of the space sources from the satellite's observations through the atmosphere, *Radiophys. Quant. Electron.*, *27*(8), 951.
155. Gurvich, A.S., and v. Kan (1997), Fluctuations of radio waves in the communication links satellite-atmosphere-satellite, *Izvestiya Russian Academy of Sciences, Physics of the Atmosphere and Ocean*, *33*(3), 314, (in Russian).
156. Wickert, J., A.G. Pavelyev, Y.A. Liou, T. Schmidt, Ch. Reigber, K. Igarashi, A.A. Pavelyev, and S.S . Matyugov (2004), Amplitude variations in GPS signals as a possible indicator of ionospheric structures, *Geophys. Res. Lett.*, *31*(24), L24801, doi:10.1029/2004GL020607, 1–4.
157. Kelley, M. C. (1989), *The Earth's Ionosphere, Int. Geophys. Ser.*, *43*, Elsevier, New York.
158. Karasawa, Y., K. Yasukawa, and M. Yamada (1985), Ionospheric scintillation measurement at 1.5 GHz in mid-latitude region, *Radio Sci.*, *20*(3), 643.
159. Hocke, K., and T. Tsuda (2001), Using GPS satellites to study plasma irregularities, *GPS World*, 34.
160. Hocke, K., K. Igarashi, M. Nakamura, P. Wilkinson, J. Wu, A. Pavelyev, and J. Wickert (2001), Global sounding of sporadic E layers by the GPS/MET radio occultation experiment, *J. Atmos. Sol. Terr. Phys.*, *63*(18), 1973.
161. Hocke, K., K. Igarashi, and A. Pavelyev (2002), Irregularities of the topside ionosphere observed by GPS/MET radio occultation, *Radio Sci.*, *37*(6), doi: 10.1029/2001RS002599, 13–1.
162. Vorob'ev, v.V., A.S. Gurvich, v. Kan, S.V. Sokolovskiy, O.V. Fedorova, and A.V. Shmakov (1999), Structure of the Ionosphere from the radio occultation GPS-"Microlab-1" satellite data: Preliminary results, *Earth Obs. Remote Sens.*, *15*, 609.
163. Pavelyev, A., T. Tsuda, K. Igarashi et al. (2003), Wave structures in the electron density profile in the ionospheric D- and E- layers observed by radio holography analysis of the GPS/MET radio occultation data, *J. Atmos. Sol. Terr. Phys.*, *65*(1), 59.
164. Pavelyev, A.G., J. Wickert, Y.-A. Liou, Ch. Reigber, T.Schmidt, A.A. Pavelyev, and S.S . Matyugov (2005), Different mechanisms of the ionospheric influence on GPS occultation signals, *GPS Solutions*, *9*(2), 96–104, doi: 10.1007/s10291-005-0138-6.
165. Pavelyev, A. G., J. Wickert, T. Schmidt, Ch. Reigber, S. S . Matyugov, A. A. Pavelyev, Y. A. Liou, and K. Igarashi (2005), Effect of solar activity in late October 2003 on radio occultation signals from the CHAMP German satellite, *Geomagn. Aeron.*, *45*(1), 134.
166. Wu, D.L., C.O. Ao, G.A. Hajj, M. de la Torre Juarez, and A.J . Mannucci (2005), Sporadic E morphology from GPS-CHAMP radio occultation, *J. Geophys. Res.*, *110* (A01066), 1–18.

167. Yeh, K.C., and C.H. Liu (1982), Radio wave scintillations in the ionosphere, *Proc. IEEE, 70*(4), 324.
168. Pavelyev, A.G., and A.I. Kucherjavenkov (1978), Radio waves scattering from the surface of a planet with the spherical symmetric atmosphere, *Radio Eng. and Electron. Phys.*, 23(8), 1569.
169. Akindinov, v.V., v.L. Naryshkin, and A.M. Ryazantsev (1976), Conductivity of the sea water, *Radio Eng. and Electron. Phys.*, 21(5), 913.
170. Pavelyev, D.A., J. Wickert, A.G. Pavelyev, I.L. Kucherjavenkova, S.S. Matyugov, K. Hocke, G. Beyerle, O.I. Yakovlev, and A. Kucherjavenkov (2003), Observation of reflected Signals in MIR/GEO and GPS/MET Radio Occultation Missions, In *First CHAMP Mission Results for Gravity, Magnetic and Atmospheric Studies Ed.*, Reigber, Ch.; Luhr, H.; Schwintzer, P. Berlin-Heidelberg, Springer-Verlag, 508.
171. Van vleck, I.H. (1947), The absorption of microwaves by uncondensed water vapor, *Phys. Rev., 71*(7), 425.
172. Barrett, A.H., and v.K. Chung (1962), A method for determination of high-altitude water-vapor abundance from ground based microwave observations, *J. Geophys. Res., 67*(9), 4259.
173. Gevakin, S.A., and A.P. Naumov (1963), Absorption of centimeter and millimeter waves in atmospheric water vapor, *Izv Vyssh Uchebn Zaved. Radiophyz, 6*(4), 674, (in Russian).
174. Gevakin, S.A., and A.P. Naumov (1964), Absorption of centimeter and millimeter radio waves in the atmospheric water vapor, *Radiotehnica i electronica, 9*(8), 1327, (in Russian).
175. Aganbek'an, K.A., A.Yu. Zraghevskiy, M.A. Kolosov, and A.V. Sokolov (1971), Investigation of dependence of absorption on the air pressure, *Radiotehnica i electronica, 16*(9), 1564, (in Russian).
176. Zraghevskiy, A.Yu. (1976), Method of evaluation of absorption in the atmospheric water vapor in millimeter and submillimeter range, *Radio Eng. and Electron. Phys., 21*(5), 951.
177. Zraghevskiy, A.Yu., and I.A. Iskhakov (1978), The total absorption of radio waves in the atmospheric water vapor in 0.8-20 mm range, *Radio Eng. and Electron. Phys., 23*(7), 1338.
178. Liebe, H., G. Hufford, and T . Manabe (1991), A model for the complex permittivity of water at frequencies below 1 THz, *Int. J. Infrared and millimeter waves, 12*(7), 659.
179. Van vleck, I.H., and v.F. Weisskopf (1945), On the shape of collision-broadend lines, *Rev. Phys., 17*(2–3), 227.
180. Van vleck, I.H. (1947), The absorption of microwaves by oxygen, *Phys. Rev., 71*(7), 413.
181. Meeks, M.L., and A.E. Lilley (1963), The microwave spectrum of oxygen in the earth's atmosphere, *J. Geophys. Res., 68*(6), 1683.
182. Naumov, A.P. (1965), The millimeter absorption spectrum of molecular oxygen, *Izv Vyssh Uchebn Zaved. Radiophyz, 8*(4), 668, (in Russian).
183. Gevakin, S.A., and A.P. Naumov (1965), Evaluation of absorption coefficient of centimeter and millimeter radio waves in the atmospheric oxygen, *Radiotehnica i electronica, 10*(6), 987, (in Russian).
184. Wulfsberg, K.N. (1967), Atmospheric attenuation at millimeter wavelengths, *Radio Sci., 2*(3), 319.

185. Rober, E.E., R.L . Mitchell, and C.I. Carter (1970), Attenuation of the 5-mm wavelength band in a variable atmosphere, *IEEE Trans. Ant. and Prop., AP-18*(4), 472.
186. Vlasov, A.A., E.N. Kadygrov, and A.N. Shaposhnikov (1990), Model for evaluation of absorption coeffeicient in the oxygen for determination of the atmospheric temperature profile from microwaves satellite's measurements, *Investigation of Earth from space*, (1), 36.
187. Gevakin, S.A., and v.S. Troitskiy (1959), Absorption of centimeter waves in layered atmosphere, *Radiotehnica i electronica*, *4*(1), 21, (in Russian).
188. Gevakin, S.A., and A.P. Naumov (1966), Absorption of electromagnetic waves in the range 3-7 mm in the earth's atmosphere, *Izv Vyssh Uchebn Zaved. Radiophyz*, *9*(3), 433, (in Russian).
189. Jacobs, E., and I.M. Stacey (1974), Atmospheric attenuation models for satellite-to earth, *IEEE Trans. Aerospace and Electron. Syst.*, *10*(1), 144.
190. Naumov, A.P., and M.B. Zinicheva (1980), Determination of vertical absorption of radio waves in the earth's atmosphere in the microwaves range transparence window, *Radio Eng. and Electron. Phys.*, *25*(5), 919.
191. Sokolov, A.V. (1974), Propagation of millimeter and submillimeter radio waves, *Itogi nauki i techniki. Radiotechnica*, *5*, vINITI Ed. (In Russian).
192. Matyugov, S.S., O.I. Yakovlev, and I.A. Vilkov (1994), Absorption of centimeter and decimeter radio waves from data of occultation experiments in communication links satellite-satellite, *J. Commun. Techn. Electron.*, *39*(8–9), 1251.
193. Eliseev, S.D., and O.I. Yakovlev (1989), Radio occultation investigation of the earth's atmosphere in the range of millimeter radio waves, *Radiophys. Quant. Electron.*, *32*(1), 3.
194. Yakovlev, O.I., and S.S. Matyugov (2005), Analysis of the atmospheric absorption of centimetric radio waves along satellite-to-satellite occultation paths, *J. Commun. Techn. Electron.*, *50*(3), 255.
195. Argenti, F., F. Cuccoli, L. Facheris, and E . Martini (2006), Tropospheric water vapor from LEO-LEO occultation: Estimation by differential attenuation measurements near 20 GHz, in the book *Atmosphere and climate studies by occultation method*, Foelsche U., Kirchengast G., Steiner A. (Eds.) Springer-Verlag.
196. Stepanenko, v.D., G.G. Schukin, L.P. Bobulev, and S.Yu . Matrosov (1987), *Radio thermal location in meteorology*, 279 pp., Gidromet Ed., Moscow, (in Russian).
197. Akvilonova, A.B., and B.G. Kutuza (1978), Radio emission of clouds, *Radio Eng. Electron. Phys.*, *23*(9), 1792.